Regeneration in Lower Vertebrates and Invertebrates: III.

Papers by
Victor B. Eichler, J. C. Streett, P. F. A.
Maderson, Lowell E. Davis, G. Kass-Simon,
S. Datta, Timothy P. Fitzharris, Mary E.
Clark, Edwin P. Marks, Gerald N. Smith, Jr.,
Barry M. Heatfield, Margaret Slaughter, et al.

MSS Information Corporation
655 Madison Avenue, New York, N.Y. 10021

Library of Congress Cataloging in Publication Data
Main entry under title:

Regeneration in lower vertebrates and invertebrates.

1. Regeneration (Biology). I. Eichler, Victor B.
QH499.R37   592'.03'1   72-8949
ISBN 0-8422-7052-3 (v. 3)

TABLE OF CONTENTS

CREDITS AND ACKNOWLEDGMENTS

Clark, Mary E., "Later Stages of Regeneration in the Polychaete, *Nephtys*," *The Journal of Morphology*, 1968, 124:483-510.

Datta, S.; and A. Chakrabarty, "Effects of Actinomycin D on the Distal End Regeneration in *Hydra vulgaris* Pallas," *Experientia*, 1970, 26:855-856.

Davis, Lowell E., "Cell Division during Dedifferentiation and Redifferentiation in the Regenerating Isolated Gastrodermis of *Hydra*," *Experimental Cell Research*, 1970, 60:127-132.

Eichler, Victor B., "Pineal Regeneration in the Frog, *Rana pipiens*, Following Embryonic Extirpation," *The Journal of Morphology, 1968, 125:253-258.*

Fitzharris, Timothy P.; and Georgia E. Lesh, "Gut and Nerve-cord Interaction in Sabellid Regeneration," *Journal of Embryology and Experimental Morphology*, 1969, 22:279-293.

Heatfield, Barry M., "Origin of Calcified Tissue in Regenerating Spines of the Sea Urchin, *Strongylocentrotus purpuratus* (Stimpson): A Quantitative Radioatuographic Study with Tritiated Thymidine," *The Journal os Experimental Zoology*, 1971, 1971, 178:233-246.

Kass-Simon, G.; and Mary Potter, "Arrested Regeneration in the Budding Region of *Hydra* as a Result of Abundant Feeding," *Developmental Biology*, 1971, 24:363-378.

Maderson, P. F. A., "The Regeneration of Caudal Epidermal Specializations in *Lygodactylus picturatus keniensis* (Gekkonidae, Lacertilia)," *The Journal of Morphology*, 1971, 134:467-478.

Marks, E. P., "Effects of Ecdysterone on the Deposition of Cockroach Cuticle *in Vitro*," *The Biological Bulletin*, 1972, 142:293-301.

Marks, E. P., "Regenerating Tissues of the Cockroach *Leucophaea maderae*: Effects of Humoral Stimulation *in Vitro*" *General and Comparative Endocrinology*, 1968, 11:31-42.

Marks, E. P.; and R. A. Leopold, "Cockroach Leg Regeneration: Effects of Ecdysterone *in Vitro*," *Science*, 1970, 167:61-62.

Marks, E. P.; and R. A. Leopold, "Deposition of Cuticular Substances *in Vitro* by Leg Regenerates from the Cockroach, *Leucophaea maderae* (F.)," *The Biological Bulletin*, 1971, 140:73-83.

Slaughter, Margaret; Florence C. Rose; and Anthony Liuzzi, "The Effect of Cesium-137 Gamma Rays on Regeneration in *Tubularia*," *The Biological Bulletin*, 1970, 138:194-199.

Smith, Gerald N., Jr., "Regeneration in the Sea Cucumber *Leptosynapta*: I. The Process of Regeneration," *The Journal of Experimental Zoology*, 1971, 177:319-330.

Smith, Gerald N., Jr., "Regeneration in the Sea Cucumber *Leptosynapta*: II. The Regenerative Capacity," *The Journal of Experimental Zoology*, 1971, 177:331-342.

Streett, J. C., "Bile Duct Regeneration in Frogs," *The Journal of Pathology and Bacteriology*, 1967, 94:459-460.

PREFACE

Regeneration of limbs in amphibians and reptiles, regeneration of eye (Woffian regeneration) and regeneration of the basic body plan in Hydra are among the classic systems for the study of differentiation and growth.

Experimental investigation of these systems has proceeded at a high rate within the last five years, as papers in the present collection illustrate. These volumes, confined to the metazoa, consider the fundamental questions of morphogenesis including the possible storage and utilization of undifferentiated cells in the adult form, the means by which cells appear to "know where they are" in developing tissue, the process of dedifferentiation, and the control of regeneration by neural and endocrine factors. Intracellular regeneration, as in the ciliate protozoa, regeneration in plants, and regeneration in planaria including the Turbellarian flatworms are covered in separate volumes.

**Regeneration of Glandular Structures
In Lower Vertebrates**

Pineal Regeneration in the Frog, *Rana pipiens,* following Embryonic Extirpation

VICTOR B. EICHLER

Information on the regeneration capacity of the epiphysis is lacking in anuran forms, although Kelly ('58) described the regeneration of normal appearing pineal vesicles after removal of epiphyses in embryonic stages of the newt, *Taricha torosa.* In an early experiment by Adler ('14) the epiphyses of *Rana temporaria* larvae were removed by thermocautery in an attempt to establish a relationship between the epiphysis and the thyroid gland. Since the animals which survived his operations failed to metamorphose, he concluded that this was evidence in support for the relationship; however, the unhealthy state of the few surviving animals may well have contributed to their restricted development. It was therefore decided that the animals used in the present study should be allowed to metamorphose, if possible, in order to test the former findings of Adler.

The frontal organ (Stirnorgan, Stieda's organ) is a nest of cells located in the dermis between the eyes of late larvae and metamorphosed *Rana pipiens.* Although several investigators (Holmgren, '18; Oksche, '52) have described a secretory function for the frontal organ, the presence of numerous photoreceptor-like cells in larval (Eakin, '61; Eakin and Westfall, '61) and adult (Oksche and von Harnack, '62; Kelly and Smith, '64) frog frontal organs supports the view that the frontal organ is a light-sensitive sensory tissue. Recordings of afferent nervous impulses (Dodt and Heerd, '62) along the nerve fibers which join the frontal organ and the epiphysis after photic stimulation within the visual spectrum have been cited as additional evidence for photoreceptor functioning of the frontal organ. However, other investigators (Stebbins, Steyn, and Peers, '60) have not found evidence to indicate either a sensory or a secretory function for this organ.

MATERIALS AND METHODS

All animals used in this study were obtained from eggs provided by induced ovulation of one *Rana pipiens* female. Pinealectomy and sham-operations were performed on embryos of Shumway stages 22–23 which were anesthetized in a weak solution of ethyl urethane. The pineal rudiment was removed with hair loop and fine glass needles through a flap made in the ectoderm overlying the rudiment. In the sham-operated animals the pineal was exposed through a similar flap in the ectoderm, but was left intact. Unoperated control animals were maintained along with the experimental animals.

Fifty pinealectomized animals and 50 sham-operated animals were used in this experiment. One day after the operations five pinealectomized animals were selected at random from the operated group for histological examination to ascertain whether all pineal tissue had been cleanly removed and to be sure that the diencephalic roof had not been unduly damaged. The examination was favorable in both respects in

[1] This investigation was supported by N.I.H. Predoctoral Fellowship GM-32771-01.

each of the five animals. The remaining 45 pinealectomized animals, and the 50 sham-operated animals were maintained throughout metamorphosis.

RESULTS

Twenty-nine (64%) of the pinealectomized animals which completed metamorphosis had regenerated some pineal tissue, although the mass of pineal tissue was not as great as that in the controls. Seventeen (58%) of the animals which had regenerated some pineal tissue also showed frontal organ development, ranging from a small clump of cells in the appropriate region to the fully formed "nest" which is characteristic of the structure. Each of these was overlain by a pigment-free clearing in the epidermis, the so-called "brow spot." Sixteen (36%) of the pinealectomized animals showed neither pineal tissue upon histological examination nor frontal organ development. In the absence of the frontal organ, no "brow spot" develops in the epidermis. The conditions of the sham-operated animals were identical to the unoperated controls in all aspects studied.

Figure 1a shows the normal condition with the "brow spot" appearing in the epidermis between the eyes. However, this animal was pinealectomized as an embryo, and had regenerated the pineal, as well as the frontal organ, as shown in figures 1b and 1c. These figures resemble the condition found in the control animals. The animal shown in figure 2a was pinealectomized at the same embryonic stage as the animal in figure 1, but no pineal tissue was regenerated. Figure 2b illustrates the histological appearance of the region of the frontal organ in this animal, and it is evident that both the frontal organ and the overlying "brow spot" are absent.

In no case was there any incidence of scoliosis (lateral spinal curvature) in the control or experimental animals in this study. Kelly ('58) described the regeneration of a normal appearing pineal vesicle after gross removal of the diencephalic roof in tailbud stages of the newt, *Taricha torosa*, and noted incidences of scoliosis in some pinealectomized animals. He suspected that there is not necessarily a causal relationship between pineal deficiency and scoliosis since damage to other brain parts was likely, due to the nature of the broad "wedge" extirpations. The proportion of pineal regeneration in non-scoliotic post-metamorphic newts, however, was greater than the proportion of regeneration reported in the present study in post-metamorphic frogs. This difference in proportion may be due to (1) the earlier stage of Kelly's animals at operation, (2) the greater pineal "field" at that stage, or (3) the possible difference in the ability of certain urodele and anuran pineals to regenerate.

DISCUSSION

In contrast to Adler's report ('14), the data obtained in the current investigation support the contention that *Rana* larvae which are pinealectomized at embryonic stages are able to grow and develop through metamorphosis into froglets with no significant departure in size from the normal condition. The failure of pineal extirpation to influence amphibian metamorphosis is found in both anurans and urodeles, since it has been reported by Kelly ('58) that larvae of the urodele *Taricha torosa*, permanently deprived of a pineal body, can develop normally well beyond metamorphosis.

Eakin ('67) reports that there was no frontal organs regeneration in the Pacific tree frog, *Hyla regilla*, after this structure was removed at larval stages. It is not known, however, whether pineal regeneration would be found in larvae of this species if they were pinealectomized at embryonic stages. Cintron and Kelly ('61) have shown that potential for pineal regeneration is present as late as stage 40 of *Taricha torosa* following discrete pinealectomy. Thus the competence of the diencephalic roof to respond to pineal removal may depend on the stage at which pinealectomy is performed, and the developmental stage through which this competence remains may differ among the urodele and anuran species which have been examined.

ACKNOWLEDGMENTS

Appreciation is extended to Dr. Jerry J. Kollros, Professor of Zoology at the University of Iowa, and to Dr. Douglas E.

Kelly, Associate Professor of Biological Structure at the University of Washington School of Medicine for the critical reading of this manuscript, and for suggestions, and to Dr. Richard G. Kessel, Associate Professor of Zoology at the University of Iowa for the use of the laboratory and photographic equipment.

LITERATURE CITED

Adler, L. 1914 Metamorphosestudien an Batrachierlarven. C. Exstirpation der Epiphyse. Roux's Arch. f. Entwmech., 40: 18–32.

Cintron, C., and D. E. Kelly 1961 An analysis of pineal regeneration in developing newts. Am. Zool., 1: 144.

Dodt, E., and E. Heerd 1962 Mode of action of pineal nerve fibers in frogs. J. Neurophysiol., 25: 405–429.

Eakin, R. M. 1961 Photoreceptors in the amphibian frontal organ. Proc. nat. Acad. Sci., 47: 1084–1088.

——— 1967 Personal communication.

Eakin, R. M., and J. A. Westfall 1961 Photoreceptors in the stirnorgan of tadpoles of the tree-frog, Hyla regilla. Am. Zool., 1: 352.

Holmgren, N. 1918 Zur Kenntnis der Parietalorgane von Rana temporaria. Ark. Zool., 11: 1–13.

Kelly, D. E. 1958 Embryonic and larval epiphysectomy in the salamander, Taricha torosa, and observations on scoliosis. J. Morph., 103: 503–537.

Kelly, D. E., and S. W. Smith 1964 Fine structure of the pineal organs of the adult frog, Rana pipiens. J. Cell Biol., 22: 653–674.

Oksche, A. 1952 Der Feinbau des Organon frontale bei Rana temporaria und seine funktionelle Bedeutung. Morph. Jahrb., 92: 123–167.

Oksche, A., and M. von Harnack 1962 Elektronenmikroskopische Untersuchungen am Stirnorgan (Frontalorgan, Epiphysenendblase) von Rana temporaria und Rana esculenta. Naturwissenshaften, 49: 429–430.

Stebbins, R. C., W. Steyn, and C. Peers 1960 Results of stirnorganectomy in tadpoles of the African ranid frog, Pyxicephalus delalandi. Herpetol., 16: 261–275.

PLATE

PLATE 1

EXPLANATION OF FIGURES

1 Animal pinealectomized at Shumway stage 22, but had regenerated
  pineal tissue. (a) Dorsal view of the head, showing "brow spot" between the eyes (arrow). Marker = 0.3 mm. (b) Transverse-section
  through the "brow spot" and frontal organ of the same animal,
  illustrating how the former is an elevated pigment-free region in
  the epidermis, and the latter is a "nest" of cells, with lumen, in
  the corium (dermis). Marker = $5.0\,\mu$. (c) Enlarged view of b.
  Marker = $2.0\,\mu$.

2 Animal pinealectomized at Shumway stage 22. (a) Dorsal view of
  the head, showing absence of "brow spot." Marker = 0.4 mm.
  (b) Transverse-section through head skin in the region corresponding to that shown in figure 1b. Marker = $6.0\,\mu$. Note absence of
  "brow spot" and frontal organ.

14

PLATE 1

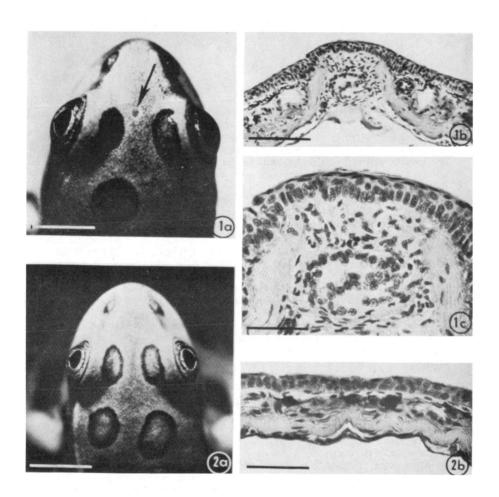

BILE DUCT REGENERATION IN FROGS

J. C. STREETT

PLATE CXL

THESE experiments were suggested by Panahandeh's (1962) report of liver damage in *Rana fusca* after ligation and division of the common bile duct. Most of his frogs died within 10 days after surgery, only one surviving for 45 days. It has since been shown by Streett (1964) that after the hepatoduodenal ligament of *R. pipiens* has been constricted by a tight ligature, the bile duct is able to by-pass the ligature so that the patency of the bile passage is restored. It was determined to repeat Panahandeh's experiments and in addition to test further the regenerative powers of the duct by removing a segment.

MATERIALS AND METHODS

The subjects were adult *R. pipiens*, mostly females, ranging from 26·8 to 78·7 g. body weight. The body cavity was opened under ether anaesthesia with routine cleanliness but not strict asepsis. The bile duct was visualised by pressure on the gall-bladder, which forced a small amount of bile to the duodenum. Between the point of its emergence from the pancreas and its insertion into the duodenum the bile duct was doubly ligated and divided between ligatures (series 1 = Panahandeh's experiment), or a segment of duct 1–5 mm. in length was removed but no ligature was applied (series 2). When the frogs were killed the patency of the bile duct was tested by pressure on the gall-bladder, and this test was then confirmed by serial sectioning.

RESULTS

Of the 19 animals in series 1, 6 out of 12 examined at 16–28 days after operation and all of 7 examined 30–60 days after operation showed regeneration of the common bile duct; of 23 animals in series 2, 9 out of 15 examined at 29–33 days after operation and 5 out of 8 examined 37–39 days after operation showed similar regeneration of the duct.

Microscopic examination of series-1 animals showed the ligatures lying within the bile-duct lumen or in a cavity communicating with it, and the presence of areas where the bile-duct epithelium was incomplete (cf. Streett, fig. 3). In series-2 animals the regenerating bile duct showed multiple short branches at the free end of its hepatic stump or at the point of junction with its duodenal stump. There were again gaps in the epithelium of the patent duct, but not so extensive as those in series-1 animals.

Since mitoses were frequently encountered in the regenerating bile-duct epithelium a survey was made, by repeating the series-2 experiments, to determine the mitotic index at its greatest and least intensities. Frogs killed 5 days after surgery showed from 3·8 to 22·5 mitoses per thousand cells compared with ⌒·14 to 1·47 in frogs 1 day after operation.

Panahandeh apparently succeeded in refuting earlier claims that tying the bile duct had no damaging effect on the liver. In *R. pipiens* also, changes in the liver architecture are frequently associated with failure of the bile duct to restore patency, but in this species it is doubtful whether 10 days' time is sufficient for such manifestations to be unmistakable. Although the possibility of a species difference is suggested, it is obvious that Panahandeh's frogs did not survive long enough to

Plate CXL

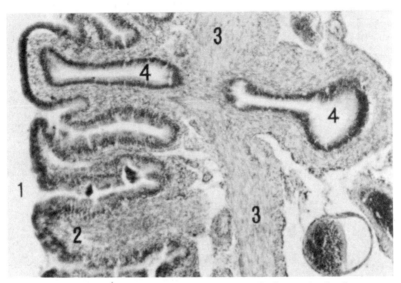

Fig. 1.—Untreated frog: normal approach of the bile duct to the duodenum.

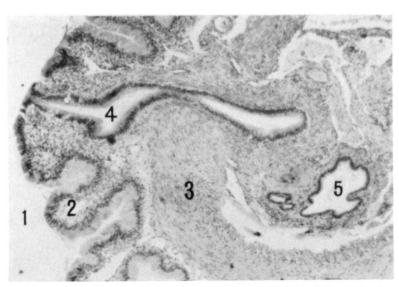

Fig. 2.—Thirty-seven days after ligation of common bile duct, patency restored. 5 = Part of a blind branch of the restored bile duct, very characteristic in such cases. A smaller branch appears above it.

Figs. 1 and 2.—Cross-sections of the duodenum at the bile duct insertion. Haematoxylin and eosin. ×75. 1 = Lumen of intestine; 2 = tunica mucosa; 3 = tunica muscularis; 4 = bile duct.

determine this point with certainty; his high mortality rate, apparently 100 per cent., is surprising, but possibly due to a difference in methods. His impression that his frogs died because they gave up eating would have no application to *R. pipiens*, which can survive prolonged starvation.

In the latter species the impossibility of predicting what will happen to the duct in any particular case suggests that whether the bile-duct stumps fuse or fail to fuse is somewhat fortuitous. In view of the tendency of regenerating epithelia to spread over surfaces (cf. Chiakulas and Millman's account, 1959, of regenerating gall-bladder in larval amphibia), one might expect such a spread from the ends of the divided bile duct, but if the development of connective tissue interrupted the surface, reunion of the stumps might well be prevented.

Restoration of patency after bile-duct ligation in rats has been reported by Trams and Symeonidis (1957), Cameron and Hou (1962), and Cameron and Prasad (1960).

SUMMARY

In 42 frogs the common bile duct was either doubly ligated and divided between ligatures (series 1, 19 animals), or had a segment removed (series 2, 23 animals). Patency of the bile duct was restored in 13 cases in series 1, and 14 in series 2.

REFERENCES

CAMERON, G. R., AND HOU, C. T. . 1962. This *Journal*, **83**, 275.
CAMERON, G. R., AND PRASAD, 1960. This *Journal*, **80**, 127.
  L. B. M.
CHIAKULAS, J. J., AND MILLMAN, M. 1959. *Anat. Rec.*, **133**, 129.
PANAHANDEH, J. . . . . . 1962. This *Journal*, **83**, 566.
STREETT, J. C., JR . . . . . 1964. *Science*, **143**, 592.
TRAMS, E. G., AND SYMEONIDIS, A. 1957. *Amer. J. Path.*, **33**, 13.

The Regeneration of Caudal Epidermal Specializations in *Lygodactylus picturatus keniensis* (Gekkonidae, Lacertilia)

P. F. A. MADERSON

The tails of lizards, like the rest of the body, are covered with scales. Each scale has outer and inner surfaces and hinge regions (Maderson, '64a). The epidermal covering of these scales undergoes periodic shedding involving the formation and loss of "epidermal generations" (Maderson, '67; Maderson and Licht, '67). It has long been known that the lizard epidermis possesses a considerable capacity for the formation of specialized structures (see review, Maderson, '70), but the exact relation of these structures to the basic epidermal generation has only recently begun to be elucidated. Many species of the gekkonine genus *Lygodactylus* possess three distinct types of specialization in their caudal epidermis. First, ventral, terminal groups of lamellar scales have a pilose surface (Loveridge, '47) similar to the digital climbing pads of other gekkonids (Maderson, '64b, '70) and anoles (Lillywhite and Maderson, '68). Second, so-called "sense organs," first described by Schmidt ('20), but only recently shown to be innervated (Miller and Kasahara, '67), are found widely distributed on tail scales. Third, in males, holocrine secretory units termed "β-glands," previously described on the posterior abdominal scales (Maderson, '68) are seen.

From both gross anatomical and microscopic viewpoints, the scales on regenerated lizard tails are the most perfectly replaced units in what has otherwise been described as a "jerry-built" structure (Woodland, '20). Minor variations in the gross form and pigmentation of the regenerated caudal integument are well known (Noble and Bradley, '33; Noble and Clausen, '36), but in some forms, e.g., pygopods (Kluge, personal communication), the regeneration of the integument is so perfect that only X-ray examination reveals the presence of a regenerated tail. Usually it is only the gross form of regenerated scales which varies from the original, and in all genera so far studied, normal epidermal differentiation patterns begin to be reestablished by the fourth week after amputation (Maderson and Roth, '70).

In *Lygodactylus*, and in a number of other lizards, not only is the normal cycle of epidermal turnover associated with shedding reestablished on the regenerated tail, but also regional epidermal specializations are reformed. The present observations describe some aspects of integumentary regeneration in *Lygodactylus* and these data are discussed in the light of current knowledge of lizard epidermal differentiation and regenerative phenomena in general.

MATERIALS AND METHODS

About 120 specimens of *Lygodactylus picturatus keniensis* (Parker) collected in northern Kenya were examined. Many specimens had either lost their tails post-mortem, or were too tightly coiled on them-selves, or were too small for easy examina-tion. They are all deposited in the collections of the Muscum of Comparative Zoology at Harvard University within the series 40924–40974: 97037–97112: 102132–102142. Among 70 specimens suitable for critical examination, 36 (18 males and 18 females) had regenerated tails, and 34 (19 males and 15 females) had intact tails. The terminal 3–7 mm of six full regenerates, two early regenerates and 11 normal tails were prepared for histological examination as follows. The specimens were decalcified in 1% nitric acid in 70% alcohol for 24 hours. The specimens were dehydrated in alcohol, cleared in chloroform, and embedded in 56°C paraffin. Serial sections were cut at 7 mμ. Some series were stained with Harris' haematoxylin and eosin, others with Harris' haematoxylin counterstained with Heiden-hein's aniline blue/orange G.

RESULTS

A. *Gross appearance*

Loveridge ('47) and Pasteur ('64) have provided detailed descriptions of the scala-tion of both original and regenerated tails of *Lygodactylus* spp., and the major fea-tures are illustrated in figure 1. On original tails the scales on the dorso-lateral surfaces are small and lozenge-shaped and overlap approximately one third of the next pos-terior scale. Under a binocular microscope minute hair-like structures may some-times be detected on the scale apices. Laterally and ventrolaterally the scales are somewhat wider than they are long, and there is a single row of much enlarged scales along the mid-ventral line (fig. 1a). This general pattern is seen along the entire length of the tail, the most anterior elements being larger than the more pos-terior. The longitudinal scale rows alter-nate with one another and cursory ex-amination reveals a distinctly ordered

pattern. At the extreme distal, ventral tip there are two rows of lamellar scales (figs. 1b,c). The pilose surface of these scales may be observed under low magnifications and the series of scales form a pad-like unit on the ventral tail tip (figs. 1b,c).

The scalation of regenerated tails differs from that of original tails, and wherever the break occurred there is a sharp dis-continuity (fig. 1a) from a regular to a haphazard arrangement. While the general trend of smallest scales in the dorsal mid-line grading into larger scales laterally, with large scales on the ventral mid-line is maintained, the ordered patterning is lost (fig. 1d). Scales with a pilose surface still form a pad-like unit on the ventral tail-tip, but the number of units involved, and their individual shapes, are greatly variable from one animal to another (figs. 1e,f).

B. *Microscopic structure*

General. Scales on all surfaces of the intact and regenerated tails show some form of the basic overlapping unit (Mader-son, '64a) with an epidermal covering over a dermal core (figs. 2, 3). In vertical sec-tion, no indication of the departure from regular patterning in regenerated tails can be detected. The more dorsal scales show a greater number of dermal melanophores than the ventral units, but the dermis has no other special features.

The epidermal covering of the scales shows the typical regional variation in the thickness of the β-layer of the outer epi-dermal generation (figs. 2–7) (Maderson, '64a). It is much better developed on the outer surface than on the inner surface and in the hinge region. The latter is fur-ther characterized by the scalloped con-volutions of its surface (figs. 2, 3, 6, 7, 9). Under oil immersion the typical spinulate gekkonid *Oberhäutchen* (Maderson, '70) may be seen over the entire epidermal surface (figs. 5, 8, 9).

The basic epidermal structure described above is interrupted by three distinct types of specialization, depending on the region of the tail being examined, and the sex of the animal. No detectable differences were found in the structure and distribution of these units in regenerated as distinct from original tails.

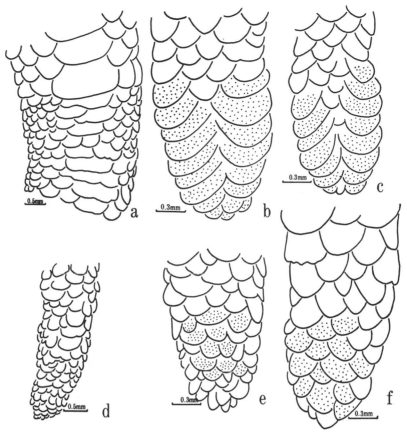

Fig. 1 Camera lucida drawings of the scalation patterns of intact and regenerated tails of *Lygodactylus*. a. — ventral view showing the sharp discontinuity between normal scalation (top) and the regenerated scalation (bottom): note that the regenerated portion shows some indications of wider scales in the mid-ventral line. b. and c. — ventral, distal tips of two intact tails showing the lamellar scales forming the scansorial "pad": scales with a macroscopically discernible "pilose" surface are stippled. d. — terminal portion of a regenerated tail in ventral view showing loss of ordered patterning. e. and f. — ventral terminal scalation of two regenerated tails showing pilose scales which form a much more haphazard form of "pad" compared to that shown in b. and c. Drawings a. to e. taken from Museum of Comparative Zoology, Harvard University, specimens catalogued as 40935, 40966, 40937, 40935, 40955, 40960 respectively.

Sense organs. The spinulate *Oberhäutchen* over the posterior tip of the dorsal scales (figs. 2, 3) (and occasionally on some lateral units) is interrupted by the presence of pairs of "hair-like" structures approximately 12 mμ in length. The underlying β-layer is very much thinner beneath these hairs, but the rest of the generation is not apparently altered. In renewal phase material, it can be seen that these hairs are modified *Oberhäutchen* cells (figs. 4, 5) (Maderson and Licht, '67). The overlying clear layer cell(s) is larger than those interdigitating with the unmodified *Ober-*

häutchen cells, and there may be some elaboration of the lacunar cells to accommodate the increased length of the specialized derivatives.

β-glands. In males, the ventro-lateral and medio-ventral scales from the cloaca to just anterior of the distal tail tip, are modified. Approximately two-fifths of their outer surfaces show some basophilic material lying between the spinulate *Oberhäutchen* and the chromophobic *β*-layer of the outer epidermal generation. The spinulate *Oberhäutchen* is interrupted at the distal extreme of this basophilic material. Renewal phase material shows that between the *Oberhäutchen* and presumptive *β*-cells of the inner epidermal generation there are one to three layers of cells of a different type (figs. 6, 7, 8). At the distal extremity of this band of tissue, there is no *Oberhäutchen* or overlying clear layer, so that the material meets the lacunar tissue directly, and the splitting plane at shedding passes between these "extra" cells and the lacunar tissue of the outer epidermal generation (fig. 8). Although these epidermal specializations are not as strongly developed as those on the pre-anal and femoral regions, especially since there is no conspicuous concavity of the dermis, they resemble exactly the basic form of the *β*-glands (Maderson, '68). They are not present in females.

Pilose pads. The distal half of the outer surfaces of the ventral terminal scales bears setae 55 μm in length, arranged in groups of four. Vertical sections of renewal phase material show a discrete row of specialized *Oberhäutchen* cells. These produce extensions on their outer surface of the same length as the mature setae, which at this time interdigitate with clear layer cells of increased depth (figs. 2, 3, 9). On the major pilose scales, there is an observable sharp discontinuity between *Oberhäutchen* cells which are specialized and adjacent unspecialized spinulate cells (fig. 9). Some scales lying proximal to the major units have somewhat enlarged spinules on their distal tips producing mature units similar to those described for *Anolis carolinensis* (Lillywhite and Maderson, '68). The positioning of the groups of specialized *Oberhäutchen* cells which produce the setae exactly resembles

that seen on the foot-pads (Maderson, '70), and there are no "mistakes" seen on the regenerated scales.

Early regeneration. The two obviously early regenerated tails (cone-like outgrowths with no external indication of scalation) were both too early to reveal the initial differentiation of the specializations described above. Both showed a thickened wound epidermis, not showing distinct epidermal generations but with the very first signs of a cephalo-caudal sequence of scale formation (Bryant and Bellairs, '67).

Abbreviations

ao, α layer of the outer epidermal generation
βi, β-layer of the inner epidermal generation
βo, β-layer of the outer epidermal generation
clo, clear layer of the outer epidermal generation
D, dermis
gc, immature glandular cells in β-glands
H, hinge region
ISS, inner scale surface
lto, lacunar tissue of the outer epidermal generation
mi, mesos layer of the inner epidermal generation
mo, mesos layer of the outer epidermal generation
Obi, *Oberhäutchen* of the inner epidermal generation
S, setae (modified *Oberhäutchen* spinules) on pilose pads
sg, stratum germinativum

Fig. 2   Low power photomicrograph of a sagittal section through an intact tail with the epidermis showing a "Stage Four" condition of the renewal phase (Maderson and Licht, '67; Maderson, '70). The dorsal scales are narrow and attenuated and bear sense organs on their apices. A region comparable to that circled and marked "4" is shown in figure 4. The developing setae (S) can be seen on the outer scale surface of the ventral scales. Dermal core (D). Apart from the fragment marked *β*o, in this and figures 3–9, the *β*-layer of the outer epidermal generation has separated from the underlying tissues during preparation and does not appear in the photographs. Hematoxylin and eosin. Magnification × 400.

Fig. 3   Low power photomicrograph of a sagittal section through a regenerated tail with the epidermis showing a "Stage Five" condition of the renewal phase (Maderson and Licht, '67; Maderson, '70). The scalation pattern appears similar to that seen in the intact tail (fig. 2), with all scales showing recognizable outer (OSS) and inner (ISS) surfaces and hinge regions (H). A dorsal apical sense organ is circled and is shown in figure 5, and a part of the regenerated pilose scale, circled, is shown in figure 9. Hematoxylin and eosin. Magnification × 400.

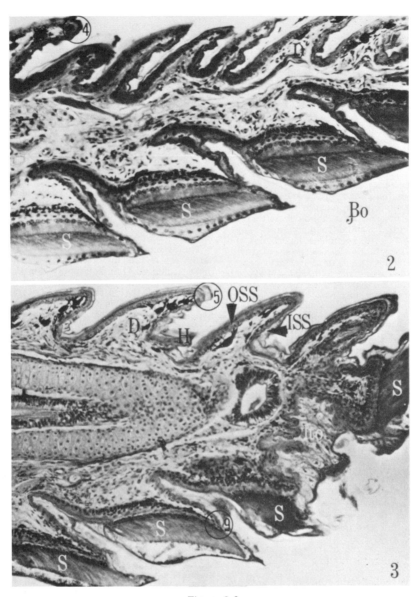

Figures 2–3

DISCUSSION

Studies of cutaneous regeneration in lizards must consider three distinct levels of organization. First, we must see whether scales are replaced and consider the extent to which they resemble the original units in gross form. Second, the capacity of the new epidermis to show generation formation must be examined, especially with reference to the quantitative variation, e.g., the relative thicknesses of the β- and α-layers, which permits identification of outer and inner surfaces and hinge regions. Third, regional specializations, their correct topographical position on the scales, and maintenance of sexual specificity where appropriate, must be identified.

The present observations show that the regenerated caudal integument of *Lygodactylus* closely resembles the original, and a similar situation has been observed with caudal pilose pads in the scincid genera *Chlorohaemus* and *Scincella* (Maderson, '69, and unpublished observations). In the gekkonid *Teratoscincus*, nearly *perfect* caudal integumentary regeneration occurs (Werner, '67). There appears to be a possible functional interpretation of these phenomena that is of relevance to the problems of cutaneous regeneration in amniotes in general, and this can now be discussed.

During mammalian cutaneous wound-healing, while the fundamental epidermal morphology is always restored, regional specializations are generally absent (Schmidt, '68). However, the nakedness of the scarred areas of the body in, for example, wild carnivores, does not appear to affect the life of the organism. Similarly, the bodies of snakes and lizards may bear large scarred regions with no apparent deleterious effects (Maderson, unpublished observations). I suggest that the functional roles of the amniote epidermis can be divided into two groups which can be correlated with regenerative capacities.

The primary role of the epidermis is that of a barrier facilitating internal homeostasis and preventing the entry of pathogenic organisms from the external environment. In any given organism the epidermis will show a basic structure which has evolved to cope with these functions. This basic structure *must* be repaired if the

animal is to survive its injury, and it follows that if the original primary barrier function is perfectly replaced, then this must be the result of the existence of a perfect repair mechanism.

A wide range of secondary functions characterizes the amniote integument e.g., thermoregulation, holocrine secretion, or sensory perception. Each of these secondary functions is associated with the presence of regional specializations, the morphological and physiological characteristics of which result from particular developmental pathways manifested during particular periods of embryonic development. Since pilo-sebaceous units or sensory areas are rarely replaced during adult mammalian wound-healing, these developmental capacities must be lost or suppressed in post-embryonic life.

The secondary functions are not necessarily *per se* less important than the primary functions, but apparently within a certain range of wound size, they can be assumed by the surrounding non-traumatized tissues. This follows the general hypothesis that regenerative capacities may

Fig. 4 Modified *Oberhäutchen* cell (x) of the inner epidermal generation forming a sense organ on an intact tail (cf. fig. 2). Hematoxylin and eosin. × 900.

Fig. 5 Modified *Oberhäutchen* cell (x) of the inner epidermal generation forming a sense organ on a regenerated tail. The structure of the sense organ, and that of the adjacent, unspecialized epidermis, is quite indistinguishable from that of an intact tail. Hematoxylin and aniline blue/orange G. × 900.

Fig. 6 β-gland on the anterior, ventral caudal scales of a male. Note the glandular cells (gc) lying between the *Oberhäutchen* and the β-cells of the inner epidermal generation (Maderson, '68). The *Oberhäutchen* is discontinuous above the glandular cells in two places: such a condition may be seen on both intact tails (as in this specimen), or there may be only a single discontinuity as in the regenerate shown in figure 7. Hematoxylin and eosin. × 600.

Fig. 7 β-gland on a regenerated tail. Hematoxylin and eosin. × 600.

Fig. 8 High power view of the *Oberhäutchen* discontinuity above glandular cells (gc) in a β-gland on a regenerated tail. The spinules of adjacent *Oberhäutchen* cells may be clearly seen. Hematoxylin and aniline blue/orange G. × 900.

Fig. 9 High power view of the distal tip of a lamellar, pilose scale on a regenerated tail (cf. fig. 3). Note the sharp discontinuity between those *Oberhäutchen* cells which give rise to modified setae (S) and those which form the normal spinules (arrow). Hematoxylin and eosin. × 900.

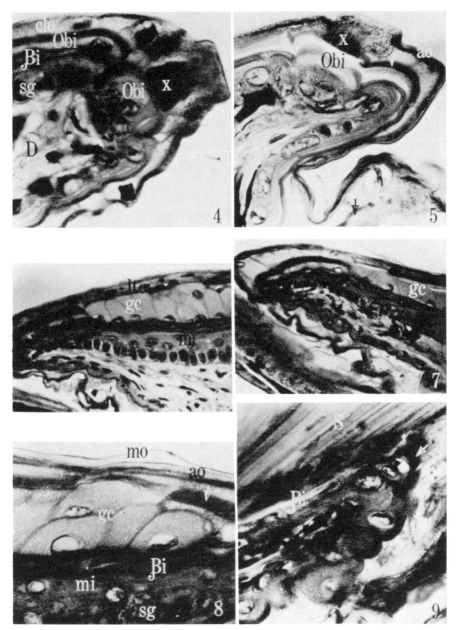

Figures 4–9

be correlated with functional demands (Goss, '69), the corollary of which is that when exceptional capacities are observed, one should be able to detect associated "exceptional" functional demands. The outstanding example of this is the complex epidermal differentiation associated with antler growth (Goss, '64). The hairs perform a vital sensory function permitting the stag to avoid contact with the environment, thus preventing damage to the delicate tissues, and also perhaps thermally insulate the highly vascular tissue.

Greer ('67, p. 7) discussed the function of the pilose caudal scales in *Lygodactylus*, and while he favors a sensory over a locomotor function (personal communication), both would be advantageous for an arboreal animal. The location of the apical sense organs on dorsal caudal scales suggests that they could function as tactile proprioceptor units (Miller and Kasahara, '67) providing information on the intensity of interscale contact which would vary with the shape of the tail arc during substrate exploration (Greer, '67, p. 7). Maderson ('70) suggested that lizard epidermal glands produce an odoriferous secretion which is spread over the substrate during locomotion, a function which could only be served from the ventral surface of the body or tail. The regenerated specializations in *Lygodactylus* are indistinguishable from the normal units. With respect to *basic* scale structure and the topographical position on the tail of the new scales bearing specializations (sense organs concentrated apically on dorsal scales, pilose pads on the outer surface of the terminal ventral scales, and β-glands on the outer surface of the ventral scales in males), cutaneous regeneration is perfect. The only "fault" is in the lack of specific resemblance of the shape of the new scales to the original. However, this does not impede the functioning of the specializations. This contrasts with the situation in the gecko *Teratoscincus* which produces a shrill, cricket-like noise with its tail (Gadow, '01; Mertens, '46; Werner, '67; Maderson and Gans, '67). During rapid sinusoidal movements, the wide, attenuated, greatly overlapping dorsal caudal scales (see figs. 1, 2, 4, and 7, Werner, '67) move over one another. The noise is produced when the series of rows of minute tuberculate derivatives of the *Oberhäutchen* cells on the distal inner scale surface move laterally across a group of similarly modified *Oberhäutchen* cells in the middle of the outer surface of the posterior-lying scale. If the mechanism is to function on the regenerated tail, not only must the specific locations of the modified *Oberhäutchen* cells be accurately restored but also the exact shape of the entire scales. The gross anatomy of this system was first described by Werner ('65) who also noted (p. 122) that the ventral scalation did not exactly resemble the original.

Hair follicle neogenesis has been discussed by Straile ('69). While it certainly occurs during antler growth (Goss, '64) and during ear-skin healing in the rabbit (Joseph and Townsend, '61), some authors believe that new follicles derive either from cellular remnants left behind following trauma or else from cells which migrate in from cells which migrate in from peripheral undamaged follicles.

Lacertilian epidermal specializations should be regarded as developmental *analogues* of mammalian pilo-sebaceous units, sensory areas, or sweat glands (Maderson, '69). Fundamentally, all of these structures, and feathers and uropygial glands in birds, are simply regions of a continuous epidermal germinal layer whose daughter cells undergo specialized patterns of differentiation (Maderson, '70). There is evidence that lacertilian epidermal specializations appear during embryogenesis (Lange, '31; Cole, '66; Maderson, unpublished), but whether they form, or indeed whether they are maintained in the adult, by the complex dermo-epidermal interactions which have been demonstrated elsewhere (Cohen, '69) is unknown. The suggested developmental relationships between various types of amniote epidermal specialization are diagrammed in figure 10.

In tail regeneration, specializations appear *de novo* within a general epidermal cell population which presumably derives from the epidermis of the proximal non-damaged tissue (Cox, '69). Possible factors influencing this development should be considered. Even it if could be demonstrated that they develop under the in-

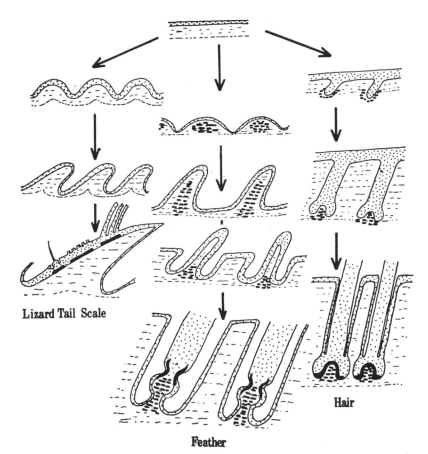

Lizard Tail Scale

Hair

Feather

Fig. 10 Diagrammatic representation of the patterns of embryonic development of a lizard scale (with associated specializations) compared with that of an avian feather and a mammalian hair. The earliest stages of differentiation of the three systems (center, top) are indistinguishable. During the formation of lizard scales, there are no signs of any dermal condensations, and the specializations appear within the epidermis of the outer scale surface when the first embryonic epidermal generation is formed. Specializations of different kinds, or groups of similar structures, are separated from one another by a "normal epidermis." Feather and hair development involve interaction of epidermal and dermal elements which are morphologically recognizable. However, the mature units are separated from one another by a "normal epidermis." Light stipple, epidermal cell populations; heavy lines within light stipple, mother cell populations of specialized epidermal structures; light dashes, dermal cell populations; dark dashes, dermal papillae of feathers and hair.

fluence of subjacent dermis, how does this tissue regain its specific inductive properties during regeneration? Differentiation of the β-glands might be influenced by a gradient effect emanating from the major pre-anal region (Maderson, '68) and sex hormones might be involved (Maderson, '70). It might be suspected that sense organs would be influenced by outgrowing peripheral sensory nerves. However, sense organs differentiate on both intact body skin and regenerating caudal tissue from *Anolis carolinensis* grown *in vitro* (Maderson, '69, and unpublished).

The outstanding feature of sense organs, and particularly the pilose pads, is the absolute specificity of positioning of the specialized cells on the outer scale surface, and the abrupt discontinuity with adjacent unspecialized cells (see figs. 4, 5, 9). These features are seen in the new units, a fact which implies that the control mechanisms involved during regeneration are the same as, or at least equivalent to, those involved during embryogenesis. Since there is absolutely no possibility that the new pilose pads could arise from undamaged or previously existing specialized cell populations, the developmental problem appears to be greater than that of hair follicle neogenesis (see above). While a capacity for tail regeneration appears to be a widespread, and probably ancient characteristic in lizards, many tissues are usually replaced to only a limited degree, and the skin is generally no exception (Noble and Bradley, '33; Noble and Claussen, '36). However, in those forms discussed here I suggest that the functional demands for restoration of the specialized caudal epidermal units is such that selection has produced a form of regenerative event previously unknown in any amniote. Impressive as regeneration of antler epidermis or rat liver may be, in both cases there is continuity with a previously existing population of specialized cells. In caudal epidermal regeneration in the lizards described here we seen an ability to return to a pattern of differentiation of specialized structures which morphologically exactly parallels the original embryonic events.

ACKNOWLEDGMENTS

I wish to thank Dr. Ernest Williams of the Harvard Museum of Comparative Zoology for making the specimens available to me and Mr. Alan Greer for his comments on the behavior of the genus. Mr. Albert Zucker made many of the histological preparations, and financial assistance for the study derived from N.I.H. grant CA-10844.

LITERATURE CITED

Bryant, S. V. and A. d'A. Bellairs 1967 Tail regeneration in the lizards *Anguis fragilis* and *Lacerta dugesii*. J. Linn. Soc. (Zool.), *46:* 297–305.

Cohen, J. 1969 Dermis, epidermis and dermal papilla interacting. In: Advances in Biology of Skin. Vol. IX. Hair Growth. W. Montagna and R. L. Dobson, eds. Pergamon Press, New York, pp. 1–18.

Cole, C. J. 1966 Femoral glands of the lizard *Crotaphytus collaris*. J. Morph., *118:* 119–136.

Cox, P. G. 1969 Some aspects of tail regeneration in the lizard *Anolis carolinensis*. I. Description based on histology and autoradiography. J. Exp. Zool., *171:* 127–150.

Gadow, H. 1901 Amphibia and Reptiles. In: The Cambridge Natural History. Vol. VIII. S. F. Harmer and A. E. Shipley, eds. MacMillan and Co., Ltd., London, 668 pp.

Goss, R. J. 1964 The role of skin in antler regeneration. In: Advances in the Biology of Skin. Vol. 5. Wound Healing. Pergamon Press, London and New York, pp. 194–207.

———— 1969 Principles of Regeneration. Academic Press, New York and London, 287 pp.

Greer, A. 1967 Observations on the behavior and ecology of two sympatric *Lygodactylus* geckos. Breviora, *268:* 1–19.

Joseph, J., and F. J. Townsend 1961 The healing of defects in immobile skin in rabbits. Brit. J. Surg., *48:* 557–564.

Lange, B. 1931 2. Integument der Sauropsiden. In: Handbuch der Vergleichenenden Anatomie der Wirbelthiere. edt. Bolk *1:* 375–447. Urban and Schwarzenberg, Berlin and Wien.

Lillywhite, H., and P. F. A. Maderson 1968 Histological changes in the epidermis of the sub-digital lamellae of *Anolis carolinensis* during the shedding cycle. J. Morph., *124:* 1–23.

Loveridge, A. 1947 Revision of the African lizards of the family Gekkonidae. Bull. Mus. Comp. Zool., *98:* 3–469.

Maderson, P. F. A. 1964a The skin of snakes and lizards. Br. J. Herpet., *3:* 151–154.

———— 1964b Keratinised epidermal derivatives as an aid to climbing in Gekkonid lizards. Nature, *203:* 780–781.

———— 1967 The histology of the escutcheon scales of *Gonatodes* (Gekkonidae) with a comment on the squamate sloughing cycle. Copeia, *1967:* 743–752.

———— 1968 The epidermal glands of *Lygodactylus* (Gekkonidae, Lacertilia). Breviora, *288:* 1–35.

———— 1969 The regeneration of epidermal specializations on lizard caudal scales. Am. Zool., *9:* 1146 (Abstract).

———— 1970 Lizard glands and lizard hands: models for evolutionary study. Forma et Functio, *3:* 179–204.

Maderson, P. F. A., and C. Gans 1967 Some observations on integumentary sound producing mechanisms in the eublepharine gekkonid *Teratoscincus* and other squamates. Am. Zool., *7:* 776 (Abstract).

28

Maderson, P. F. A., and P. Licht 1967 Epidermal morphology and sloughing frequency in normal and prolactin-treated *Anolis carolinensis* (Iguanidae, Lacertilia). J. Morph., *123:* 157–172.

Maderson, P. F. A., and S. I. Roth 1970 Observations on integumentary regeneration in lizards. Am. Zool., *10:* 556 (Abstract).

Mertens, R. 1946 Die Warn- und Droh-Reaktionen der Reptilien. Abh. senckenb. naturf. Ges., *471:* 1–108.

Miller, M., and M. Kasahara 1967 Studies on the cutaneous innervation of lizards. Proc. Calif. Acad. Sci., *34:* 549–568.

Noble, G. K., and H. T. Bradley 1933 The effect of temperature on the scale form of regenerated lizard skin. J. Exp. Zool., *65:* 1–16.

Noble, G. K., and H. J. Clausen 1936 Factors controlling the form and color of scales on the regenerated tails of lizards. J. Exp. Zool., *73:* 209–229.

Pasteur, G. 1964 Rechèrches sur l'évolution des Lygodactyles – lézards Afro-malgaches actuels. Trav. Inst. Scient. Cherif (Zool.), *29:* 5–132.

Schmidt, A. J. 1968 Cellular biology of vertebrate regeneration and repair. University of Chicago Press, Chicago and London, 420 pp.

Schmidt, W. J. 1920 Einiges über die Hautsinnesorgane der Agamiden, insbesondere von *Calotes,* nebst Bemerkungen über diese Organe bei Geckoniden und Iguaniden. Anat. Anz., *53:* 113–139.

Straile, W. E. 1969 Dermal-epithelial interaction in sensory hair follicles. In: Advances in Biology of Skin. Vol. IX. Hair Growth. W. Montagna and R. L. Dobson, eds. Pergamon, London and New York, pp. 369–392.

Werner, Y. L. 1967 Regeneration of specialized scales in *Teratoscincus* (Reptilia: Gekkonidae). Senck. Biol., *48:* 117–124.

Woodland, W. N. F. 1920 Some observations on caudal autotomy and regeneration in the gecko (*Hemidactylus flaviridis* Ruppel), with notes on the tails of *Sphenodon* and *Pygopus.* Q. J. Micr. Sci., *65:* 63–100.

Regeneration In Hydra

CELL DIVISION DURING DEDIFFERENTIATION AND REDIFFERENTIATION IN THE REGENERATING ISOLATED GASTRODERMIS OF *HYDRA*

LOWELL E. DAVIS

Previous studies on the regenerating isolated gastrodermis have shown that this single layer of *Hydra* is capable of regenerating into a complete animal [4, 5]. Furthermore, animals derived from the isolated gastrodermis were apparently normal, and formed buds as well as gonads [2]. The ultrastructural studies indicated that gland cells located at the periphery of the regenerating explant (gastrodermis) dedifferentiated into interstitial cells and subsequently redifferentiated into cnidoblasts [4], nerves (Davis, in preparation) and sperms [2]. The digestive cells at the periphery of the explant transformed directly into epithelio-muscular cells without an intervening interstitial cell stage [4].

Several questions have arisen concerning cell division, both in the case of the interstitial cells derived from dedifferentiated gland cells and epithelio-muscular cells derived from the transformation of digestive cells. It was noted earlier [4] that although interstitial cells undergo division in the normal animal, no mitoses were observed in the system. The appearance, however, of nests of up to sixteen interstitial cells suggested that division of these cells had occurred. Epithelio-muscular cells are also known to undergo division in the normal animal [1], but division of these cells was not observed in the earlier investigations [4].

The present study shows division of interstitial cells derived from dedifferentiated gland cells in the isolated gastrodermis, division of epithelio-muscular cells that have transformed from digestive cells, and the unusual division of cnidoblasts (see also [3]). These cnidoblasts arose from interstitial cells which had dedifferentiated from gland cells.

MATERIALS AND METHODS

The animals used in these experiments were *Hydra viridis*. They were cultured by the method of Loomis & Lenhoff [6] except that distilled water was substituted for tap water. Normal, non-budding animals were starved for 24 h prior to the isolation of the gastrodermis. Explants of the gastrodermis consisting of gland cells and digestive cells only were obtained and cultured by the method described previously [2, 4, 5]. The technique used is extremely reliable in obtaining gastrodermis free of epidermal cell types.

Whole pieces of regenerating tissues were fixed in glutaraldehyde by a modified method of Sabatini et al. [7] and post-fixed in osmium tetroxide. After fixation the tissues were dehydrated rapidly in graded concentrations of alcohol and embedded in Maraglas. Sections were cut with glass and diamond knives on a Cambridge ultramicrotome and mounted on carbon-coated Formvar-filmed grids. Sections were stained in uranyl acetate and lead citrate. They were examined in a RCA EMU 4 electron microscope.

OBSERVATIONS

In an earlier report on the dedifferentiation of gland cells into interstitial cells, it was noted that the newly derived interstitial cells formed nests of up to 16 cells as are seen in the normal animal. Although no interstitial cells were observed in actual division, it was assumed that division occurred. Fig. 1 shows one interstitial cell in division and another in interphase. It is not possible in these studies to determine whether the interstitial cell in interphase is a newly derived cell from a gland cell or one of the daughter cells from a previous division. Also, we have not determined how many times the interstitial cells divide prior to differentiation into other cell types.

The differentiation of interstitial cells into cnidoblasts has been well documented by several investigators. However, the division of cnidoblasts themselves is an unusual phenomenon. Fig. 2 shows two cnidoblasts undergoing nuclear division. Their dividing nuclei, centriole and spindle fibers are similar to those of other cells in normal mitosis. In both cells, the plane of section passes through the developing nematocysts. The cytoplasm contains many ribosomes, small mitochondria and large segments of granular endoplasmic reticulum.

Transformation of digestive cells into epithelio-muscular cells occurs during the regeneration of the isolated gastrodermis. Although it was assumed in earlier studies that the resulting epithelio-muscular cells divided, none of these cells were observed in division. However, since epithelio-muscular cells are known to undergo division in the normal animal, the above assumption seemed reasonable. Fig. 3 shows an epithelio-muscular cell in an early stage of mitosis. The chromatin material is dispersed throughout the nucleus, and the nuclear membrane is ruptured in several areas (*arrows*). Small mucous droplets, microtubules, vacuoles are of various sizes and muscle fibers are present as well as the usual cell organelles.

DISCUSSION

Division of several cell types occurs in non-sexual *Hydra* [1]. In one instance, division of interstitial cells provides a supply of undifferentiated cells from which specialized cells develop (e.g., cnidoblasts). In another instance, differentiated cells such as epithelio-muscular cells, digestive cells and mucous cells divide in order to replenish their respective cell types.

If growth must occur during the regeneration of the isolated gastrodermis into complete animals, some degree of cell division is necessary. The present study shows that division of at least some interstitial cells does occur, some subsequently differentiating into other cell types.

It would be interesting to determine the exact time that DNA synthesis occurs prior to the division of the three cell types. For example, when does DNA synthesis occur during the dedifferentiation of the gland cell and the formation of the interstitial cell? Also when does DNA synthesis occur and what is the fate of the cnidoblasts (derived from the interstitial cells above) that are in the process of synthesizing their nematocysts

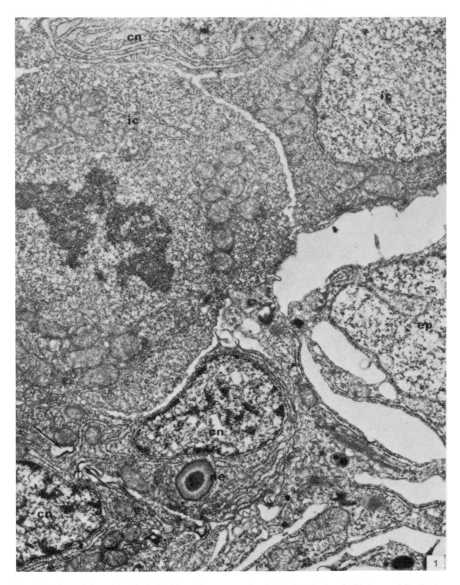

Fig. 1. Portion of the epidermis derived from the regenerating isolated gastrodermis. Note especially the two interstitial cells (*ic*), one of which is in division. Portions of three cnidoblasts (*cn*) are also observed, two of them being connected by an intercellular bridge (*arrows*). A nematocyst (*ne*) is seen in one cnidoblast. Part of an epithelio-muscular cell (*ep*) is also seen. × 17,000.

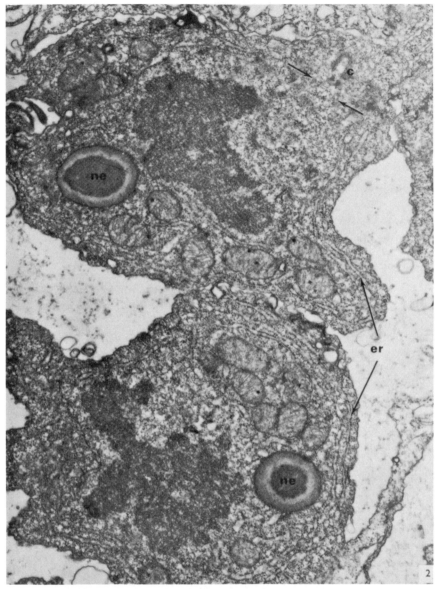

Fig. 2. Two cnidoblasts undergoing nuclear division. The nematocysts (*ne*), mitochondria and endoplasmic reticulum (*er*) are located toward the periphery of the cells. A centriole (*c*) and spindle fibers (*arrows*) are seen in one cell (*top*). × 22,000.

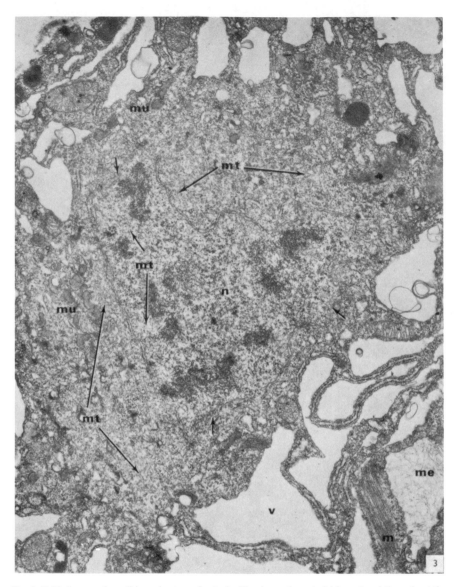

Fig. 3. Epithelio-muscular cell in early stage of mitosis. The chromatin material is scattered throughout the nucleus (*n*), and the nuclear membrane is ruptured in several areas (*arrows*). This cell extends from the external mucous border (*extreme top-left*) where there are mature mucous droplets to the mesoglea (*me*). Note the small mucous droplets (*mu*), microtubules (*mt*), vacuoles (*v*), and muscle fibers (*m*) at the base of the cell (continuation between muscle fibers and the remainder of the cell is not shown). × 17,000.

36

and at the same time undergoing nuclear division? Preliminary autoradiographic studies (with tritiated thymidine) are inconclusive.

Further investigations are currently underway in an attempt to resolve these questions and other related problems.

The author wishes to thank Professor Allison L. Burnett, Department of Biology, Northwestern University, Evanston, Ill. for his criticisms and review of the manuscript. The technical assistance of Mrs Gloria Curatolo is gratefully acknowledged.

This investigation was supported by the National Science Foundation, grant GB-8384.

REFERENCES

1. Burnett, A L, Am naturalist 100 (1966) 165.
2. Burnett, A L, Davis, L E & Ruffing, F E, J morphol 120 (1966) 1.
3. Davis, L E, Exptl cell res 52 (1968) 602.
4. Davis, L E, Burnett, A L, Haynes, J F & Mumaw, V R, Develop biol 14 (1966) 307.
5. Haynes, J F & Burnett, A L, Science 142 (1963) 1481.
6. Loomis, W F & Lenhoff, H, J exptl zool 132 (1965) 555.
7. Sabatini, D D, Bensch, K & Barrnett, R J, J cell biol 17 (1963) 19.

Arrested Regeneration in the Budding Region of Hydra as a Result of Abundant Feeding

G. KASS-SIMON AND MARY POTTER

INTRODUCTION

When hydra are very well fed they produce buds near the lower portion of the gastric region, just above the peduncle. When such budding animals are transected in the vicinity of the bud, head regeneration is significantly delayed compared to nonbudding animals cut in the same region (Burnett, 1961; Kass-Simon, 1969a). Regions outside of this area, however, regenerate within the normal time (Kass-Simon, 1969a).

Current theory ascribes this delay to a diffusing inhibitor substance produced either by the differentiated "head" of the bud (Tardent, 1960; Burnett, 1961) or to the proliferating cells of the growing bud (Burnett, 1966). The causal relationships of the model may be stated in this way: Proliferating cells (or the differentiated head) produce(s) a chemical substance which inhibits further proliferation and differentiation in other, neighboring cells. The relative concentration of this substance with respect to a growth stimulating substance determines the nature of a morphogenetic event (Burnett, 1966; Lesh and Burnett, 1964). A concentration gradient resulting from changes in this ratio results in a morphogenetic hierarchy (Burnett, 1966; Corff and Burnett, 1969).

Our experiments lead us to believe that the concepts concerning the relationship between budding and regeneration ought perhaps to be modified and that it is possible to explain the antithesis between budding and regeneration without reference to an internally produced growth inhibitor. In our experiments large quantities of food prevent the budding region from responding to a wound stimulus with either head or basal plate regeneration by causing *incipient* budding in that region. We are able to show that basal plate formation can be prevented by high levels of food *without* a bud ever being produced. At the same time we can show that bud production does not preclude further bud formation. We believe that our results are most easily explained by postulating that the budding zone tissue enters a "budding state" as a result of high food consumption which renders it temporarily incapable of responding to a wound stimulus with a regeneration response.

Our experiments fall into two parts. First, we use the upper half of transected animals to show that basal plate regeneration in the budding region can be blocked by high levels of food consumption. Then we try to turn the regeneration response on and off by varying food quantity and wound number. Secondly, we allow lower halves of of very well fed animals to regenerate heads in orders to show that in this case, too, the regeneration response is preempted by incipient bud formation.

PREPARATION OF EXPERIMENTAL ANIMALS
AND EVALUATION PROCEDURES

The animals used in these experiments belong to a single asexually reproducing clone of *Hydra attenuata* Pall. They are maintained in "BVC" solution (Loomis and Lenhoff, 1956) at $20°C \pm 2°C$ and fed an excess of *Daphnia pulex* daily. The experimental animals themselves are derived in the following way: Animals bearing young buds are placed in individual petri dishes and each is hand fed one full-grown *Daphnia* per day until the bud has detached from its parent. The parent is discarded and the bud is used as the actual experimental animal. These animals are then maintained in individual petri dishes and fed according to their experimental schedules once every day. Water is changed every other day, while undigested *Daphnia* remains are removed daily with a pipette.

All transections are made on millimeter paper embedded in a paraffin-filled petri dish. Animals transected in the "budding region" are cut exactly in half, as measured from mouth to basal plate. "Wounds" are made by cutting small slivers or wedges from the previously cut edge of the animal. During the course of an experiment animals were inspected daily with a dissecting microscope (60-fold magnification) and sketched. Although basal plates are clearly discernible under this magnification, we always checked for stickiness by running the flat edge of a pair of forceps along the bottom and sides of the regenerating animal. (Peduncle formation was not always evident.)

Where applicable a χ^2 test was used to evaluate the data.

The Effect of Various Food Levels on Basal Plate Regeneration in the Budding Zone

The purpose of this experiment is 2-fold. We want to show that it is only in the budding region that high food levels block basal plate regeneration, and that the block occurs in the absence of bud formation and depends entirely on the quantity of food consumed.

Procedure. All animals were transected 5 days after they detached from their parents. They were divided into feeding groups which were subjected to different feeding schedules. All groups were transected in the budding region with the exception of the first group, which was cut above the budding zone at the upper ¼ of the total length (see Fig. 1). This group represents one of the controls and was fed two *Daphnia* per day from day of detachment until 3 days after transection. The second group of controls were cut in the budding region and were never fed from day of detachment. The third group were fed from day of detachment until day of transection: half of the group received one *Daphnia* per day, the other half, two *Daphnia* per day. The fourth group were fed two *Daphnia* per day until 3 days after transection, and the last group were fed two *Daphnia* per day until they either regenerated or were discarded.

Results. The results are summarized in Table 1 and Fig. 1. All animals cut above the budding zone regenerated within 2 days. Similarly, all unfed animals cut in the budding zone regenerated within 48 hours. Of the experimental animals, those fed for the shortest period of time, until the day of transection, regenerated within 2 days. No buds were produced during the 48 hours in which the animals regenerated. They were discarded immediately after they produced basal plates. Presumably they would have made buds had they been kept longer with continued feeding. Among the well fed experimental animals, buds were produced only after the fifth day after initial transection (see below.) At the next level of food consumption, however, the number of animals regenerating within 48 hours is significantly reduced; only 9 out of 19 animals fed two *Daphnia* per day until 3 days after detachment regenerated within 48 hours (ca. 48%). This drop is significant at $p < 0.001$ ($\chi^2 = 11.5$) in comparison to the unfed controls. One animal first produced a bud 5 days after transection, indicating that regeneration may have been blocked before actual bud production had begun. Nine other animals showed regeneration delays of from 2 weeks (1 case) to 2.5 months (1 case) (see Fig.

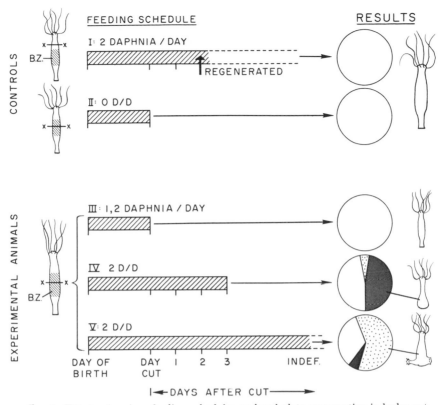

Fɪɢ. 1. Effects of various feeding schedules on basal plate regeneration in hydra cut in the budding zone. Clear sector = percentage of animals which regenerate and do not bud; black sector = percentage of animals which neither regenerate nor bud; stippled sector = percentage of animals which bud but do not regenerate; BZ = budding zone. See text for actual percentages.

2). This last animal was discarded before it ever produced a basal plate. All animals which regenerated during the first week did so within 48 hours after transection. After this, the time of regeneration was scattered over the next several weeks: two did not regenerate until the beginning of the sixth week after transection. All of these "arrested regenerates" became discernibly smaller, and none of them ever produced buds or any other differentiated structure at the transection site as far as we could tell.

When feeding was prolonged indefinitely even fewer animals regenerated. Only 6 out of 18 animals produced basal plates within 2 days (ca. 33.3%). This proportion, however, is not significantly different

TABLE 1
REGENERATION ARREST IN ANIMALS CUT IN THE BUDDING REGION AND
SUBJECTED TO VARIED CONDITIONS OF FEEDING[a]

Feeding schedule		Total number of animals transected	Number of animals rerating within 1 week after transection	Number of animals failing to regenerate within 1 week after transection	
Number of Daphnia per day	Fed from day of detachment until:			Nonbudding (arrested-regenerates)	Budding
Control animals					
2	3 Days after transection	10	10	0	0
0	Not fed	20	20	0	0
Experimental animals					
1 or 2	Day of transection	20	20	0	0
2	3 Days after transection	19	9[b]	9	1
2	Day of regeneration	18	6[b]	1	11

[a] The first group was cut at $\frac{1}{4}$ total length. All other groups are cut at $\frac{1}{2}$ total length.

[b] The proportion of these animals not regenerating within the first week is significantly larger than the controls ($\chi^2 = 11.5$ ($n = \frac{9}{19}$) and 16.5 ($n = \frac{6}{19}$), $p < 0.001$). The difference between these groups is not significant.

from the 48% which produced basal plates in the previous group, ($\chi^2 = 0.28$). There is also no difference between the ratio of nonbudding arrested regenerates to regenerates and the ratio of budding arrested regenerates to regenerates ($\chi^2 = 0.85$). The difference, however, is significant with respect to the unfed controls at $p < 0.001$ ($\chi^2 = 16.5$). Eleven of the 18 animals produced buds at the transection site at the steady rate of about one per day. Since each bud remained attached to the parent for about 3 days, it was almost always the case that a single animal bore two or even three buds simultaneously. Only rarely did one bud detach before a second was formed. Clearly then, neither the presence nor the formation of a bud prevents another bud from being produced, although basal plate formation is always precluded by bud production. Only 3 animals began to bud during the first week after transection; the others did not begin to bud until 3 or 4 weeks after cutting, an observation suggesting that the first effect of food on the budding zone is to make the tissue

Fig. 2. An arrested basal plate regenerate. This animal was transected in the budding region 2.5 weeks before being photographed. It had been fed two *Daphnia* per day until 3 days after transection.

unresponsive to a regeneration stimulus and only subsequently to cause the tissue to begin bud production. During the time between transection and budding no other differentiated structure was produced at the transection site.

The Effect of Lowering Food Levels on the Regeneration Response of Repeatedly Wounded Arrested Regenerates

The purpose of this experiment was to see whether lowering the food level of arrested regenerates would again allow them to respond to a wound stimulus by regenerating a basal plate. We also wanted to see whether changes in the food level influenced the nature of the regenerated structure.

Procedure. The animals used in this experiment were those animals which where fed two *Daphnia* a day until 3 days after transection and which did not regenerate within the first 72 hours after transection. These were termed "arrested regenerates" since if animals do not regenerate within the first 48 hours they are unlikely to

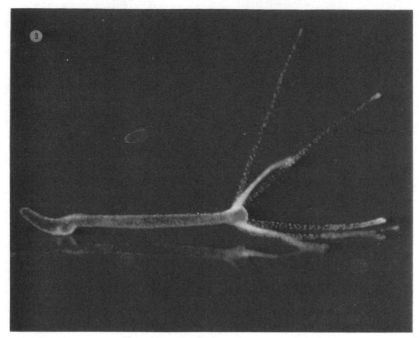

FIG. 3. Basal plate regeneration in response to a second wound. This animal was wounded 2.5 weeks after the original transection. The wound was made at some distance from the transection site; presumably the basal plate regenerated at the site of the wound. This animal is not included in the present data.

do so before the second week after transection. (see Results in section above.)

Arrested regenerates, then, are wounded for the first time 3 days after the original transection (wound No. 1). Any animal failing to regenerate a basal plate within 72 hours of this wound is again wounded at 72 hours after the first wound (wound No. 2). A third wound is given to those animals which do not regenerate within 72 hours of wound No. 2. Wound No. 4 is given to any animal failing to regenerate within 72 hours of wound No. 3. The entire series is repeated with another set of arrested regenerates; but this time the first wound is not given until 6 days after the original transection. So that in terms of absolute time, the first wound of this series corresponds to the second wound of the previous series. In the same way, the experiment is repeated a third time, but this time the first wound is not administered until 9 days after the original transection and corresponds in absolute time to wound No. 3 of the first group. Thus by

44

staggering the time at which the first wound is administered, it is possible to see whether there is any difference in the regeneration response of animals starved for 1 day and those starved for 13 days.

Results. The results are summarized in Table 2. Withdrawing food from arrested regenerates for even 1 day allows the majority of animals to become responsive to a wound stimulus. Regardless of whether the first wound was administered after 1, 4, or 7 days of starvation, $2/3$ of the total number wounded responded to the wound with basal plate regeneration within 48 hours of the wound (see Fig. 3). This 48-hour figure is exactly the same as for unfed controls in the previous experiment and appears to indicate that the wound initiates a specific program of events leading to basal plate formation. The response to subsequent wounding was also either basal plate formation within 48 hours or nothing. There was no difference in the number of animals responding to a second wound among the three experimental groups. Even when the last wounds were administered as late as 12 and 15 days after the initial transection (10 and 13 days without food), the only response to a wound was basal plate formation or nothing. In no case was any other differentiated structure produced at the wound site during the intervals between stimulus and response.

The Effect of Continuous Feeding on the Regeneration Response of Repeatedly Wounded Arrested Regenerates

This experiment represents the control for the previous experiment. Its purpose was to see whether arrested regenerates would be prevented from responding to additional wounds if food consumption was maintained at a high level.

TABLE 2

Effect of Wounding on Basal Plate Regeneration in Arrested Regenerates Where Feeding Was Stopped after Initial Transection[a]

Time of 1st wound	Number of animals regenerating within 72 hours after wound number				
	1	2	3	4	
3 Days after transection	8	2	1	1	$N = 12$
6 Days after transection	7	1	1		$N = 9$
9 Days after transection	6	2	1		$N = 9$

[a] There are 72 hours between each successive wound. The proportion of animals responding to the first wound is not significantly different among the three groups.

45

Procedure. The animals in this experiment were maintained on the same feeding schedule as the last group of experimental animals in the first experiment. That is, they were continuously fed two *Daphnia* per day from day of detachment until they either regenerated or were discarded.

Animals which are maintained on this feeding schedule and which have not regenerated basal plates within 72 hours of transection are given a first wound at 72 hours. Those failing to regenerate within 72 hours of this wound (including now those animals which have begun to produce buds) are given a second wound at 72 hours after the first wound. This is 6 days after the initial transection (see Fig. 4). Wound No. 3 is given 9 days after the initial transection to those

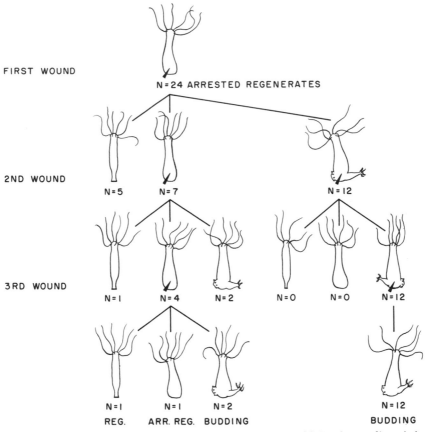

Fig. 4. Effect of continued feeding on responsiveness to additional wounding. Animals are fed 2 *Daphnia* daily and wounded every 3 days.

animals which have failed to respond to wound No. 2 within 72 hours. Wound No. 4 is given 12 days after the original transection to those animals which have not regenerated within 72 hours of wound No. 3.

Results. The results are summarized in Fig. 4. When animals were starved in the previous experiment, about ⅔ of the total wounded for the first time 3 days after transection responded by regenerating basal plates. In this experiment, under conditions of continuous feeding, significantly fewer animals respond to a first wound with basal plate regeneration. Only about ⅕ (5 out of 24) of those wounded regenerated basal plates. The difference is significant at $p < 0.02$ ($\chi^2 = 5.43$) of the 19 animals which did not regenerate after the first wound 12 began to produce buds at the transection site instead, suggesting that the tissue had already been preempted for bud formation. Seven animals remained arrested. At the second wound, one of these seven regenerated a basal plate, two others now produced buds and 4 remained arrested. When these four were again wounded, two produced buds, one remained arrested and one regenerated. Buds were always produced immediately at or very close to the transection site indicating that there had been no shift in budding zone tissue. Despite repeated woundings none of the budding animals ever regenerated basal plates within the 3 months they were kept. Buds were produced as in the first experiment at the steady rate of about one per day and it was usually the case that an animal carried more than one bud at any given time. When feeding was stopped after 3 months, all animals regenerated although day to day checks were no longer made.

The Effect of Large Food Quantities on Head Regeneration of Lower Halves of Animals Transected in the Budding Region

Since it was not possible to vary the food intake of lower regenerating halves, we were unable to produce any arrested regenerates as defined above. The purpose of citing these experiments, therefore, is to show that food also prevents head regeneration in the budding region of lower halves by causing the tissue in this region to begin bud production instead.

Procedure. The animals in this experiment were transected 24 hours after removal to individual petri dishes. Each animal was given six *Daphnia* at the time of removal and cut in the budding zone 24 hours later. Control animals are represented by animals which had been given only one *Daphnia* at the time of removal and cut in the budding zone 24 hours after feeding ($N = 16$). These animals are cited in a previous paper as part of another experiment (Kass-Simon, 1969a): All 16 animals in that case regenerated within 2 days after transection.

Buds are distinguished from regenerated heads by the fact that they always produce a basal plate even when they do not detach from their parents. Buds usually appear to grow away from the parent stalk at an obtuse angle, whereas head regenerates always begin with the formation of an apex surrounded by tentacle knobs directly at the top of the transected stalk.

Results. The results are summarized in Fig. 5. Of the 16 animals transected, eight regenerated heads within 2 days of transection. This is significantly fewer than regenerated among the controls. ($p < 0.001$, $\chi^2 = 12.3$). Eight other animals produced buds at the

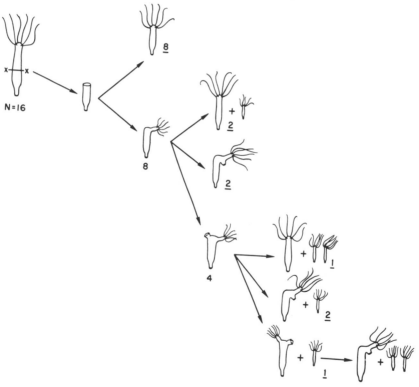

FIG. 5. Bud production by lower halves of very well fed animals transected in the budding zone. Animals which did not regenerate remained attached to their buds at the gastric region. All buds produced basal plates. See text for further explanation.

transection site *in lieu* of regenerated heads 2 days after cutting. No bud formation was visible before this time. Of these eight, two regenerated the day after the bud detached; another two remained permanently attached to the gastric region of the bud. In these cases the buds made basal plates at their sides and later formed peduncles which grew down parallel to the stalk of the mother animal. The other four animals which produced buds on the second day, produced a second set of buds on the third day. (see Fig. 6). Of these four which produced second buds, one made a third bud to which it remained permanently attached and two remained permanently attached to their second buds in the manner just described. In all cases where mother and bud remained attached to one another, a basal plate was formed indicating that the heads in this case are properly considered to be differentiates produced by the growing bud rather than regenerates produced by the mother animal. Budding and head formation were the only alternative morphogenetic events observed at the transection site.

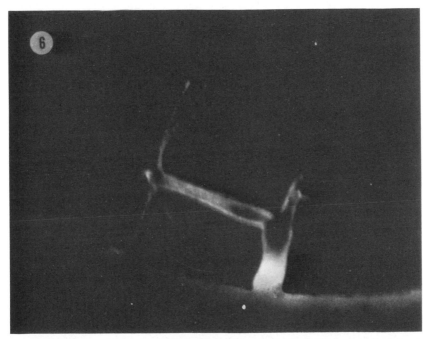

Fig. 6. Bud production at the transection site of a lower half of a heavily fed animal. Both buds eventually produced basal plates and detached from the mother animal. The mother later regenerated hypostome and tentacles.

DISCUSSION

Our experiments show that high food levels will prevent the tissue of the budding zone from regenerating basal plates without ever causing it to produce buds. In addition, we also show that the budding zone tissue remains unresponsive to wound stimuli as long as the food level remains high and that it returns to a wound-responsive condition as soon as the food level is reduced. Our results could easily be explained by saying that the reason for the inability of the budding zone tissue to respond with regeneration is that it has already been designated for bud production. This seems plausible since when food is at a very high level bud upon bud is produced at the wound site instead of a regenerate. What we are saying is that high levels of food turns off the responsiveness of the budding zone tissue at the same time it initiates the budding process.

Since this formulation differs from the currently well-known inhibitor model, it is necessary to explain why the inhibitor-gradient model does not adequately fit the results of our experiments. First, it is clear that fully differentiated buds are not necessary to inhibit regeneration, since basal plate regeneration can be blocked in the absence of any bud. Secondly since in our experiments bud formation *never* prevents the formation of more buds, but only the regeneration of basal plates, any presumed inhibitor produced by the proliferating cells must be one which prevents only the processes of regeneration and not the proliferation and differentiation of budding. This would make it inaccurate to talk of a "growth" inhibiting substance.

A corollary of the inhibitor model is that it is possible to construct a morphogenetic hierarchy which depends on the relative concentration of inhibitor to stimulator in a given tissue undergoing a morphogenetic event. The concentration of inhibitor varies directly with the rate of cell proliferation (Burnett, 1966), and it has been shown by Corff and Burnett (unpublished results, 1970) that the rate of proliferation is dependent upon food quantity. Thus we would expect that if the rate of proliferation is changed before regeneration is allowed to occur, the response to a wound would not always be the differentiation of the same structure. In our experiments we varied feeding from starvation for 2 weeks to constant feeding for 2 weeks. In all cases the response to wounding upper halves was either basal plate regeneration within 48 hours or nothing. It is true that it might still be said of this result that we were unable to vary the proliferation rate sufficiently to allow the cells of the budding zone to reach either a higher or lower rung in the morphogenetic hierarchy. But the formulation of any such hierarchy will itself lead to conceptual difficulties.

On the basis of our experiments we would have to define the hierarchy for upper halves in this way: buds, basal plates, hypostome and tentacles, since the immediate alternative to budding is basal plate regeneration. But clearly, if we do define the hierarchy in this way we are immediately faced with a dilemna. In our experiments in which lower halves are allowed to regenerate, the only alternative to budding is head formation. This would mean that the ranking would have to read: buds, hypostome and tentacles and then basal plates. [Corff and Burnett (1969) also rank basal plates below heads although they entirely omit buds.] It is obvious, therefore that if proliferating cells were producing a single inhibitor whose concentration alone determines the nature of a differentiation, its effect is different for lower halves regenerating heads than it is for upper halves making basal plates: since contradictory hierarchies must be constructed for the two cases. Except to say that body regions of hydra maintain strict polarity, this commonplace finding has not been incorporated into the currently accepted model.

Since it is difficult to fit our results into the existing inhibitor model, we would like to propose an alternative explanation which would fit our data and the findings of others without leading to any apparent contradictions. On the basis of our experiments, then, we do not question the possible existence of structure specific inhibitors produced by both the head and basal plate producing cells of *Hydra*. Nor do we wish to argue against the possibility that proliferating cells produce a local unstable *general* inhibitor to differentiation. We do, however, feel it is possible to explain the antithesis between budding and regeneration without recourse to an internally produced inhibitor substance.

As already said here and elsewhere (Kass-Simon, 1969b) our results lead us to believe that high food levels cause a radical change in the tissue of the budding region which renders it temporarily incapable of responding to a wound stimulus. The level at which this change occurs can only be guessed at. It may be that the change takes place at the level of tissue structure as suggested by the findings of Burnett and Hausman (1969; Hausman and Burnett, 1970) who reported a breakdown in the mesogleal fabric of the budding region as a result of feeding. (A similar breakdown, however, is found during regeneration.) Alternatively, the change might occur directly in the cells themselves. Possibly at the moment of wounding, food may already have caused the cells to enter a proliferative or preproliferative state in which they would not be capable of producing the appropriate mRNA and protein required for regeneration. Evidence for this kind of mechanism is given for other systems: The inhibition of mRNA and protein synthesis during mitosis is reported for several eukaryotic cells (Prescott and Bender, 1962; Konrad, 1963; Fan and Penman, 1970). It is likely too that food may act more directly on the genome in a way similar to the action of small ions. Kroeger (1963, 1964) describes sequential changes in the puffing pattern of chironomous chromosomes as a result of changes in the ratio of Na^+ and K^+ ions. It may be that a similar quantitative change in a food contained factor switches the genome from a "non-budding" (regeneration responsive) to a "budding" (regeneration unresponsive) condition. These questions remain open for the time being.

SUMMARY

Abundant feeding can cause animals that have been transected in the budding region to delay basal plate regeneration for 2 weeks to 2.5 months. These animals do not produce any buds. In the absence of food, when such animals are wounded again, their ability to regenerate basal plates within 2 days returns. On the contrary, under conditions of constant feeding, these animals do not respond to additional wounds with basal plate formation. They either remain arrested regenerates or begin bud production at the transection site.

Lower halves of heavily fed animals transected in the budding region produce buds *after* transection instead of head regenerates.

The results are interpreted to mean that large quantities of food cause the tissue of the budding region to switch over into a "budding" condition which simultaneously renders it temporarily incapable of responding to a wound stimulus.

This work was begun at the Zoologisches Institut der Universität Zürich (Professor Dr. Ernst Hadorn, Director) and completed at the laboratory of L. M. Passano, The University of Wisconsin. We are very much indebted to Dr. Passano for reading the manuscript and for his helpful and cogent criticism.

REFERENCES

BURNETT, A. L. (1961). The growth process in *Hydra*. *J. Exp. Zool.* **146**, 21–83.
BURNETT, A. L. (1966). A model of growth and cell differentiation in *Hydra*. *Amer. Naturalist* **100**, 165–189.

BURNETT, A. L. and HAUSMAN, R. E. (1969). The mesoglea of *Hydra*. II. Possible role in morphogenesis. *J. Exp. Zool.* **171**, 15–24.

CORFF, S. C., and BURNETT, A. L. (1969). Morphogenesis in *Hydra*. I. Penduncle and basal disc formation at the distal end of regenerating hydra after exposure to colchicine. *J. Embryol. Exp. Morphol.* **21**, 417–443.

FAN, H. and PENMAN, S. (1970). Regulation of protein synthesis in mammalian cells. II. Inhibition of protein synthesis at the level of initiation during mitosis. *J. Mol. Biol.* **50**, 655–670.

HAUSMAN, R. E., and BURNETT, A. L. (1970). The mesoglea of *Hydra*. III. Fibre system changes in morphogenesis. *J. Exp. Zool.* **173**, 175–186.

KASS-SIMON, G. (1969a). The regeneration gradients and the effects of budding, feeding, actinomycin and RNAse on reconstitution in *Hydra attenuata* Pall. *Rev. Suisse Zool.* **76**, 565–599.

KASS-SIMON, G. (1969b). Regeneration in the budding zone of *Hydra*: Control of the response by feeding and wounding (Abstract.) *Amer. Zoologist* **9**, 611.

KONRAD, C. G. (1963). Protein synthesis and RNA synthesis during mitosis in animal cells. *J. Cell. Biol.* **19**, 267–277.

KROEGER, H. (1963). Chemical nature of the system controlling gene activities in insect cells. *Nature (London)* **200**, 1234–35.

KROEGER, H. (1964). Zell physiologische mechanismen bei der regulation von genaktivitäten in der riesen chromosomen von *Chironomus thumi*. *Chromosoma* **15**, 36–70.

LESH, G., and BURNETT, A. L. (1964). Some biological and biochemical properties of the polarizing factor in *Hydra*. *Nature (London)* **204**, 492–493.

LOOMIS, W. F., and LENHOFF, H. M. (1956). Growth of *Hydra* in mass culture. *J. Exp. Zool.* **132**, 555–573.

PRESCOTT, D. M., and BENDER, M. A. (1962). Synthesis of RNA and protein during mitosis in mammalian tissue culture cells. *Exp. Cell. Res.* **26**, 260–268.

TARDENT, P. (1960). Principles governing the process of regeneration in hydroids. *In* Developing Cell Systems and Their Control (D. Rudnick, ed), pp. 21–43. Ronald Press, New York.

Effects of Actinomycin D on the Distal End Regeneration

in *Hydra vulgaris* Pallas

S. Datta   and A. Chakrabarty

The process of hypostome and basal disc regeneration in hydra involves activation, migration, multiplication and transformation of cells[1]. Morgan[2] has proposed epimorphosis and morphallaxis as the mechanism of restitution in hydra. This was later supported by authors like Sanyal[3] and Sanyal and Mookerjee[4]. In order to see the effects of inhibitors of energy metabolism on this dynamic process, several antimetabolites were used under different conditions without much effect on the initial regeneration of the distal end in *Hydra vulgaris*[5]. Recently Clarkson[6] studied in detail DNA, RNA and protein

[1] S. Mookerjee and S. Bhattacherjee, Wilhelm Roux Arch. EntwMech. *157*, 1 (1966).

[2] T. H. Morgan, *Regeneration* (MacMillan, New York 1901).

[3] S. Sanyal, Experientia *18*, 449 (1962).

[4] S. Sanyal and S. Mookerjee, Folia biol. *15*, 237 (1967).

[5] S. Datta, Ind. J. exp. Biol. *6*, 190 (1968).

[6] S. G. Clarkson, J. Embryol. exp. Morph. *21*, 33 (1969).

synthesis during first hour of distal end regeneration of decapitated *Hydra littoralis*. He found no appreciable increase in DNA synthesis, a large increase in RNA synthesis and a slight increse in protein synthesis. He further studied effects of Actinomycin D(AD) on RNA and protein synthesis in hypostome regenerating hydra and obtained substantial suppression of RNA synthesis without much effect on protein synthesis[7]. Hypostome formation was not inhibited completely under these conditions. In the present study AD at various concentrations applied at different hours before and after amputation was tested with a view to studying its effect on the distal end regeneration.

Material and method. Hydra vulgaris Pallas were cultured following the method of Loomis and Lenhoff[8]. 3 different experimental conditions were set up. The animals were cut at the subhypostomal level and immediately transferred to AD at 5 and 10 μg/ml in hydra solution. In the second set of experiments, amputation was done after several hours of pretreatment with AD solution and kept in the same solution for observation. In the third set of experiments the actinomycin treated animals after amputation were left in normal hydra solution. Controlled animals were maintained with each set of experiments. When 75% of the animals in each set-up regenerated, the results were considered as positive. The

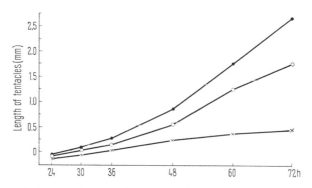

Fig. 1. Showing the rate of increase of the length of tentacles after Actinomycin-D treatment. ●—●, control; ○—○, 5 μg/ml: ×—×, 10 μg/ml.

[7] S. G. Clarkson, J. Embryol. exp. Morph. *21*, 55 (1969).
[8] W. F. Loomis and H. M. Lenhoff, J. exp. Zool. *132*, 555 (1956).

increase in length of the tentacles was considered as the rate of regeneration in this study. The length of each tentacle was measured in relaxed state on the screen of

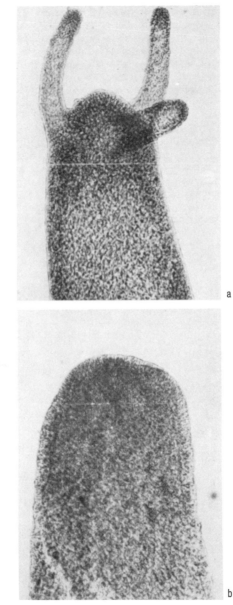

Fig. 2. a) Hydras pretreated in Actinomycin-D for 48 h regenerated after 72 h b). Pretreated for more than 48 h, failed to regenerate.

a Reichert Visopan Microscope. The mean of the total length of all the tentacles at a particular hour was divided by the magnification value to obtain the real length, and is shown in the Figure.

Results and discussion. Normally a decapitated *Hydra vulgaris* takes about 24 h to reconstitute hypostome with tentacle rudiments. When hydras were kept in 10 µg/ml AD solution immediately after amputation, the reconstitution time was delayed by an average of 6 h and the rate of regeneration appreciably decreased during the first 72 h of differentiation, after which cytolysis followed and the animals disintegrated. In 5 µg/ml series only the rate of regeneration decelerated (Figure).

Whole hydras were pretreated in 10 µg/ml AD separately for 24 and 48 h and subsequently were amputated. In the former case, a definite rate of regeneration up to 48 h was observed, followed by gradual disintegration. In the latter case, there was no differentiation at all and by 96 h all had disintegrated. Hydras treated in 5 µg/ml AD did not disintegrate till 10 days.

When 24 h pretreated animals in 10 µg/ml AD brought back to normal solution after amputation, differentiation started after 48 h. The 48 h pretreated animals differentiated after 72 h (Figure 2, a). The 72 h pretreated animals did not differentiate but continued to live for 10 days or more (Figure 2, b) and no regeneration occurred in 96 h pretreated animals and all disintegrated beyond 4 days.

Decapitated hydras treated with AD at 10 µg/ml up to 48 h could not totally suppress the initial reconstitution process, but animals pretreated more than 48 h before amputation failed to regenerate. This indicates that although there is a burst of RNA synthesis at the early hours of hypostome determination[6], it is not really essential for the initial reconstitution process. The initial differentiation of the proximal end of hydra cut at the subhypostomal level is perhaps accomplished by structural proteins, synthesized with the help of a stable variety of messenger RNA. The existence of a stable mRNA or a masked templet material for initial hypostome determination in hydra has been respectively suggested by CLARKSON[7] and DATTA[5]. A masked RNA has also been reported to be responsible for the AD resistant protein synthesis in *Arbacia* egg[9]. From the above results it was revealed that the differentiation could only be suppressed if treatment with AD was done more than 48 h before amputation. This could perhaps be the time required for total turnover of the preexisting stable mRNA associated with the initial determination process. Treatment for more than 48 h in 10 µg/ml AD possibly

[9] P. R. GROSS and G. H. COUSINEAU, Expl. Cell Res. *33*, 368 (1964).

inflicts permanent damage on the metabolic activities of the cell, as a result of which no regeneration took place[10].

[10] The authors are indebted to Prof. S. MOOKERJEE, head of the Department of Zoology, Presidency College for lending laboratory facilities.

Regeneration in Annelids, Molluscs and Insects

Gut and nerve-cord interaction
in sabellid regeneration

By TIMOTHY P. FITZHARRIS AND GEORGIA E. LESH

Studies of annelid regeneration have considered interactions which may exist between the gut and nerve-cord. Although the influence of the gut has often been assigned a secondary role (Kroeber, 1900; Morgan, 1902; Hunt, 1919; Faulkner, 1932; Sayles, 1932), the necessity for its presence has been demonstrated in certain instances (Okada, 1938). Attention therefore has been primarily directed to the nerve-cord, particularly the trophic influences of this structure at the wound site (Goldfarb, 1914; Bailey, 1930 & 1939; Avel, 1932; Crowell, 1937), and its influence posterior to the wound area, presumably hormonal in nature (Kropp, 1933; Clark, R. B. & Clark, M. E., 1959; Clark, R. B. & Bonney, 1960; Clark, M. E. & Clark, R. B., 1962; Scully, 1964; Golding, 1967 a–c). Regardless of the organism or the approach used, previous investigators have pointed to one persistent problem: the independence or interdependence of these two organ systems in the structuring of the regenerate bud. Specifically, what does the gut contribute to the regenerate bud and in what manner, if any, does the nerve-cord direct the formation of the bud? Because of the lack of histological data, plus conflicting views on the exact ordering of these regenerating organ systems, this study will attempt to resolve the problem of tissue interaction involved in normal regeneration.

A unique developing system provided by the sabellid polychaetes can be employed to clarify this situation. Sabellids are marine annelids characterized by the presence of three distinct body regions: (1) a bilobed branchial crown of tentacles used as a respiratory and feeding organ; (2) a thoracic region generally 5–11 segments in length, with *dorsal* filiform chetae and *ventral* uncinigerous hooks; and (3) an abdominal region composed of 0–300 segments with *dorsal* hooks and *ventral* chetae (Fig. 1). When the animal is cut through the thoracic or abdominal region, anterior regeneration will be epimorphic and limited (hypomeric), i.e. only the prostomium, peristomium and first thoracic segment will be replaced, regardless of the level of amputation. If less than the original number of thoracic segments remain, new thoracic segments will be replaced by a process of post-cephalic reorganization, i.e. the orientation of the

60

abdominal chetae will be reversed and the overall characteristics will become thoracic in nature. These observations are easily corroborated using a high-power dissection microscope. Posterior regeneration involves an initial 2-day wound-healing period. Subsequently the extension of the digestive tract, with commensurate growth in related regions, forms a pygidium which directs normal growth (Nicol, 1930; Berrill, 1931, 1952, 1961; Huxley & Gross, 1934; Gross & Huxley, 1935; Berrill & Mees, 1936; Herlant-Meewis, 1964).

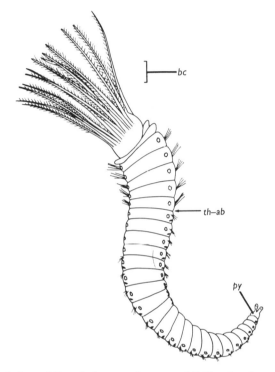

Fig. 1. Dorsal view of the whole normal worm exhibiting the three characteristic body regions. The branchial crown (*bc*) is a respiratory and feeding organ; the thoraco-abdominal junction (*th–ab*) marks the division of the two regions of the trunk while the pygidium (*py*) denotes the last segment.

Transection of the animal at any level therefore reveals the presence of three consistent structural elements—gut, ventral nerve-cord and body wall—any of which could contribute to the regenerative process. Because of the precise positioning of these structures at the wound region, their interactions will of necessity be quite limited. In this paper the mutual dependence of both gut and nerve-cord in the initiation and maintenance of regeneration will be demonstrated. Evidence in support of this will be drawn from surgical manipulation and treatment with colchicine.

The polychaetes *Sabella melanostigma* and *Branchiomma nigromaculata* are obtained commercially from Tropical Atlantic Marine Specimens, Big Pine Key, Florida. Animals are maintained in 78 and 195 l. aquaria at room temperature (22·5 ± 0·5 °C) in constantly aerated artificial sea water ('Instant Ocean', Aquarium Systems Inc., Wickliffe, Ohio). Stock cultures are fed daily with tropical fish food. Fresh tap water is added periodically to the tanks to maintain a specific gravity of 1·025.

Experimental animals are kept in 11 cm finger bowls if 1–5 are used, or in 19 cm finger bowls if more than five are cultured simultaneously. 'Instant Ocean' (IO) is changed daily and the finger bowls continuously aerated. Animals are starved during the entire experimental period. All animals are anaesthetized 10–15 min in 0·1 % chloretone for all operations and/or photographic recordings.

Surgical procedures

Experimental animals were derived from whole worms severed at the thoraco-abdominal junction. These abdominal portions were further subdivided into ten segment pieces and therefore were regenerating simultaneously anterior and posterior portions. Experimental animals regenerated individuals in the same manner and at the same rate as animals severed only anteriorly or posteriorly. Abdominal segments also showed post-cephalic reorganization (parapodial inversion) clearly and rapidly. This, plus the repetitive histological organization of the abdominal metamere, reduced experimental variation. Both *Branchiomma* and *Sabella* regenerated similarly under all conditions tested, and therefore results of any one section apply to both genera (see Table 1).

Excision of the gut at the anterior end was done by means of a lateral incision

Table 1. *A representation of the various morphological features during normal regeneration*

Time after operation (days)	Morphological features
1 2	Eversion of gut; wound healing
3	Gut resorbed; wound covered by epithelium
4 5	Regenerate buds
6	Prominent buds
7 8	Finger-like bud stage
9	Growth with elaboration of tentacles
10	Initiation of post-cephalic reorganization
11	Collar segment
12	Palps; first thoracic segment

into the body wall 3–4 segments long, with a subsequent dorsal incision to produce an epidermal flap. Bending this flap over exposed the gut, which could be removed by grasping with forceps (Fig. 2).

The ventral nerve-cord was removed similarly from the anterior end by an incision on either side of the cord. This piece of body wall plus the closely adhering nerve-cord was removed entirely (Fig. 3). Animals with identical sham incisions, plus normal regenerates, served as controls. Three segment portions were preferably removed in all experimental cases since more extensive removal (six segments—total length) of gut and/or nerve-cord resulted in the death of the animals.

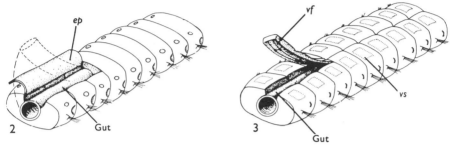

Fig. 2. Diagrammatic representation of the surgical procedure for removal of the gut (dorsal view). The epidermal flap (*ep*) bends back and exposes the gut, which is removed with forceps. By refolding the flap, wound healing is facilitated.

Fig. 3. Diagrammatic representation of the surgical procedure for removal of the nerve-cord. The ventral flap (*vf*) is bent back and removed entirely, taking that section of the nerve-cord closely adhering to it. The ventral shields (*vs*) flanking the groove denote the ventral side.

For histological investigations, animals were anaesthetized in 0·1 % chloretone, fixed in Baker's dilute Bouin's, dehydrated in an ethanol series, cleared in xylene, and embedded in paraffin wax. Sections were cut at 7 μ, mounted on glass slides, and stained in 0·1 % toluidine blue.

Colchicine studies

Stock $2·5 \times 10^{-3}$ M solutions of colchicine (Nutritional Biochemicals) were freshly prepared and stored in light-proof containers no longer than 5 days. Culture solutions were changed daily with experimental animals being removed at specific intervals to IO and cultured normally. The final concentration of $1·25 \times 10^{-3}$ M was selected from the following data:

Concentration (M)	Gross morphological effects
$2·5 \times 10^{-3}$	Death
$1·25 \times 10^{-3}$	Delayed regeneration; polarity affected
$2·5 \times 10^{-5}$ $2·5 \times 10^{-6}$ $2·5 \times 10^{-7}$	Normal regeneration

Excision of the gut

To determine the significance of the gut in anterior regeneration, twenty worms were anaesthetized and the gut was removed for three segments behind the wound area. Wound healing occurred in approximately 18 h with a notable constriction around the incision.

The flat surface of the wound area was drawn 1–2 segments into the body, expanding the lateral region of the worm while concurrently positioning itself in the direction of the severed end of the gut. Sometimes the ventral half of the body curled up to the third segment to meet the dorsal epithelium, thus effecting wound closure where the gut had been removed. In both cases, however, the result was the same—the positioning of the anterior wound surface close to the anterior portion of the intact gut. These experimental animals regenerated normally (see Table 1 for staging), with the buds invariably produced dorsally at the original amputation site, never at the point of excision of the gut (Fig. 4).

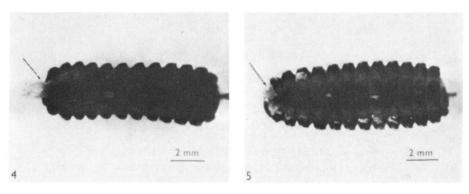

Fig. 4. Seven-day regenerate with the gut removed for three segments posterior to the wound site. Note that tentacle formation (arrow) is at the most anterior part of the animal, not at the point of excision of the gut.

Fig. 5. Seven-day regenerate with the nerve-cord removed for three segments posterior to the wound site. Note the overall distortion of the animal into an L-shape which projects the regenerate tentacles toward the reader (arrow). Inspection will reveal that the buds are on the *dorsal* side in their proper position at the wound site, not originating from the cut end of the nerve-cord.

In the normally amputated worm such a structural collapse would be limited to the first injured segment because the gut would support the constricting epidermal wall. In the experimental animals constriction of the outer wall with the ventral constriction combined to draw all the remaining structures as far posteriorly as possible; that is, to the point where the gut had not been removed. Subsequent regrowth of the gut was therefore less than the length of one total segment in order to communicate with the outside. Otherwise it would have had to regrow the length of three segments to reach the wound surface.

Histological examination of animals with the gut removed supports the gross observations and has revealed several additional points. Even though the gut has been effectively removed for three segments posterior to the wound area, the anterior part of the gut remains at the wound site. The reason for this is that wound closure after such an operation is generally accomplished by a tight constriction of the outer body wall collecting the remaining structures and forcing them toward the unoperated region like a collapsible accordion. The constriction of the muscular elements does not meet with any resistance since the gut and connected septa have been removed. Normally, constriction of these circular muscles would affect wound closure by tightly constricting around the gut and closing 90 % of the wound mechanically. Cicatrical material combined with an overgrowth of epithelium later joins the wall to the gut completing the process of closure.

Therefore, in *Sabella* and *Branchiomma* removal of the gut for three segments posteriorly does not in essence remove the gut from the wound region. The influence of this structure on the anterior regenerating end may have been reduced but certainly not eliminated. The actual positioning of the gut in relation to the regenerate bud has not been clearly demonstrated in previous investigations. Morgan (1902) relies upon stitching to effect healing and reports in some cases that a head develops where the anterior cut surface of the nerve-cord is present whether gut tissue is present or not. Hunt (1919) also attempted a description of the relationship between these structures but the lack of histological pictures makes an adequate interpretation difficult.

Excision of the nerve-cord

Removal of the nerve-cord for three segments in twenty animals was done simultaneously with the excision of the ventral groove and a small portion of the body wall. Upon healing the dorsal body wall was drawn ventrally, causing the animal to assume an L-shaped appearance. Because of this body distortion the regenerate buds appeared to be produced ventrally (Fig. 5). Microscopic observation reveals that the buds are dorsal to the gut on either side in their normal position (Fig. 6). Growth of a new cephalic region always occurs at the wound site, no supernumerary heads being produced at the anterior surface of the nerve-cord (as reported in Morgan, 1902).

Histological examination demonstrates an odd relationship between the gut and the nerve-cord under these experimental conditions. The gut is drawn over the cut end of the cord in such a way that a longitudinal section of the animal shows a cross-section of the gut (Figs. 6, 7). This relationship is a direct reflexion of the L-shaped configuration visible at the gross morphological level. Removal of the cord for three segments posteriorly is therefore without effect in the overall regeneration of the animals. The important aspect is again the method of wound closure. After such an operation there is a very tight contraction of the circular muscles in the body wall with a concomitant contraction of the longitudinal muscles. Because part of the ventral structural elements is missing, the sum total of this reaction draws the worm into an L-shaped figure at the

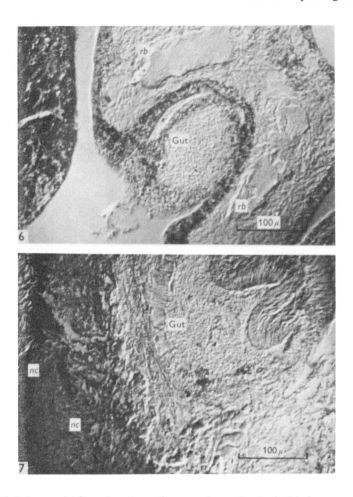

Figs. 6, 7. Sequential frontal sections of an experimental animal with the nerve-cord removed. (*rb*, Regenerate buds; *nc*, nerve-cord.) Photomicrographs of histological sections stained with toluidine blue were taken using a Normarski differential interference microscope to increase structural detail.

Fig. 6. A cross-section of the gut showing its relation to the newly formed regenerate buds.

Fig. 7. A cross-section of the gut with a longitudinal section of the nerve-cord, illustrating the distortion seen at the gross morphological level in Fig. 5. This demonstrates that although the nerve-cord was removed for three segments its influence on the regenerating region does not necessarily act at a distance.

anterior end. The histological results show that the unsupported structures anterior to the severed end of the cord are drawn ventrally and bent almost at right angles. Frontal sections of such worms show a cross-section of the gut at the point where the nerve-cord ends.

This mechanical aspect of wound healing essentially accomplishes two things, wound closure and the drawing into apposition of the gut with the nerve-cord, thus eliminating the necessity of anterior regeneration of the nerve-cord for three segments. Subsequent development shows regenerating nerve fibers innervating the region laterally around the gut in the formation of a new supra-esophageal ganglion. Thus, as in the case of the gut, excision of the nerve-cord for three segments does not preclude a role for the nerve-cord at the regenerating end.

This is the main factor overlooked in both Morgan's and Goldfarb's experiments because their major supposition is that the nerve-cord is adequately removed *spatially* from the regenerating region. Bailey (1930) and Crowell (1937) overcame this situation to some extent when they performed similar operations on *Eisenia* and *Allolobophora* respectively, by looping the nerve-cord back into the body. However, following these manipulations, Bailey reported no regeneration while Crowell demonstrated that regeneration proceeded normally. Comparable results were obtained by Okada & Kawakami (1943) and Okada & Tozawa (1944).

The combined removal of gut and nerve-cord served as an operational control. Initial regeneration in this group was considerably slowed, but bud formation and subsequent development proceeded normally. The positioning of the gut and nerve-cord in respect to each other and to the regenerate bud was completely normal.

Interdependence of the intact gut and nerve-cord

Additional evidence for mutual interplay of the gut and nerve-cord during regeneration may be demonstrated by achieving conditions in which there is either, (1) an intact gut with a severed nerve-cord, or (2) an intact nerve-cord with a severed gut. To obtain these situations twenty whole normal worms were subdivided into two groups in the following manner: worms in group 1 were bent in a U-shape so that both *ventral* surfaces of the worm were in contact and secured with cotton thread; worms in group 2 were similarly bent with their *dorsal* sides in contact and secured. A one-to-two-segment divit removed from the exposed bend in these animals with micro-dissecting scissors produced an everted gut at two surfaces with an intact nerve-cord in group 1, and an intact gut with two cut ends of the ventral cord in group 2.

The results of this study were as follows. Animals with intact nerve-cords and severed guts (group 1) healed by a tight constriction around the wound area, drawing the two ends of the severed gut toward each other, and forcing them back into their original position. Regrowth of the epidermis completed wound closure with no unusual effects observed.

Wound healing in group 2 was incomplete or absent. No tight constriction of the ventral body wall occurred to aid wound healing as was observed for the dorsal side.

Thus, an interesting aspect concerning possible mechanical differences between the dorsal and ventral body walls is noted. Healing is easily accomplished from the dorsal aspect by a tight constriction of the body wall around the ends of the everted gut, thus restoring the individual to normal. The ventrally operated region does not contract and in several cases the wound is widened by the constant writhing of the animal. Failure to close the wound results in the death of the animals.

The most prominent anatomical feature of the ventral body wall is the ventral shield, a glandular cushion on every segment, which secretes the mucus lining the tube (Nicol, 1930). This ventral shield, as compared to the dorsal body wall, is semi-rigid. This explains why the ventral wall is less pliable and poses a limiting factor in wound closure. These structural differences do not imply qualitative differences in the ability of the tissue to respond to stimulation, but rather show the limitations of the body wall as a whole in controlling nerve-cord and gut interaction.

A simple model of this idea may be constructed by using tubing of different diameter. Similar incisions produce equal surface areas at the wound region. Closer inspection, however, will reveal that the total surface area which communicates with the coelom (which must be closed) will be much less in the case of the severed gut than of the severed nerve-cord.

The fact that no supernumerary heads or tails are observed when the ventral nerve-cord is severed is surprising. It is reported that such growths can be achieved by deflextion of the nerve-cord in the sabellid *Spirographis spallanzani* (Kiortsis & Moraitou, 1965). Failure to respond adequately to the initial wounding may have prevented the nerve-cord from interacting with the surrounding tissue to elicit this response. This interaction has been demonstrated in lumbricid oligochaetes by looping the nerve-cord back through a small hole in the body wall and obtaining supernumerary structures (Avel, 1947).

Colchicine studies

One controversial point confronting any attempt to define tissue interactions during regeneration is the origin and role of migratory cells (neoblasts, coelomocytes, regeneration cells) in this process. Numerous studies have been conducted (Liebmann, 1942 *a, b*; Stephan-Dubois, 1958; Lender, 1962) but the actual contribution of these cells in forming a regenerate bud has never been explicitly demonstrated. The alkaloid colchicine is known to be an effective mitotic inhibitor (Taylor, 1965). In addition, its ability to inhibit the migration of cells during regeneration has been reported (Lehmann, 1957, 1965). Therefore, this chemical provides an approach whereby one may gain further insight into the mechanisms of reorganization at the cellular level.

To determine the role of cell migration in tissue interactions involved in annelid regeneration, animals were exposed continuously to $1·25 \times 10^{-3}$ M colchicine. Culture solutions were aerated and changed daily as in other experiments. Groups of experimental worms removed daily from colchicine to IO exhibited a delayed pattern of regeneration shown in Table 2.

Table 2. *The effect of continuous exposure to colchicine on regeneration*

(* = removed to IO; B = buds; T = tentacles; C = controls.)

Time in colchicine (days)	Time after operation (days)															
	1	2	3	4	5	6	7	8	9	10	11	12	13	14	15	16
6	—	—	—	—	—	*	—	—	—	—	B[1]	B	B	T	*T*	T
5	—	—	—	—	*	—	—	—	—	B	B	B	T	*T*	T	T
4	—	—	—	*	—	—	—	B	B	*T*	T	T	T	T	T	T
3	—	—	*	—	—	—	—	—	B	B	B	T	*T*	T	T	T
2	—	*	—	—	—	—	—	B	B	B	B	*B*	T	T	T	T
1	*	—	—	—	—	—	B	B	B	B	*T*	T	T	T	T	T
C	—	—	—	—	—[2]	B	B	B	B[2]	T	T	T	T	T	T	T

[1] Italic indicates expected time of regeneration after treatment (see note [2]).
[2] Days 1 and 5 of visible regeneration (see Table 1).

The daily lag in the removal of worms from colchicine to IO was reflected in the overall pattern of regeneration. The appearance of regeneration buds was delayed, but this delay was in direct proportion to the time cultured in colchicine. The inhibitory effect was therefore instantaneous and was maintained until the animals were returned to IO. In several instances worms were cultured in colchicine up to 2 weeks with no ill effects (i.e. they regenerated upon return to IO), thereby eliminating sublethal cytolysis as an experimental factor. Normal healing patterns (eversion of the gut with contraction of the outer body wall) were observed in all experimental animals, but there was a definite and substantial delay in the fusion of the two layers. Therefore, colchicine may be acting here to block cell division in the epithelial layer and in the other tissues as well, because an immediate overgrowth of the epithelium was evident when the animals were returned to IO. Further corroboration of these effects is currently being investigated by means of autoradiography.

When these results are compared with those on *Nephtys* and *Nereis*, an interesting hypothesis may be proposed. If regeneration in sabellids is under neurohormonal control, it is possible that colchicine affects not only cellular migration and division but also neurosecretion. If this situation is the case, two alternatives may be proposed. (1) The cells responsible for the formation of the regenerate buds may be incapable of responding while exposed to colchicine, since regeneration does not occur. Returning the experimental animals to IO initiates regeneration but on a delayed time scale. Therefore, it would have to be postulated that a stable hormone (active) is produced with effectiveness lasting at least up to 2 weeks in certain instances. (2) Alternatively, the rate of manufacture and/or release of the hormone may be critical.

Clark & Clark (1962) on *Nephtys* originally demonstrated in transplantation experiments that decerebrate animals receiving an 'activated' ganglion from a post-5-day regenerating donor do not regenerate. The assumption here was that the neurosecretory products of the implanted ganglion had already been released. Golding (1967*a–c*), using *Nereis*, however, demonstrated that ganglia from intact animals and post-5-day regenerating animals are capable of inducing regeneration in host worms. By doing sequential transplants of the same ganglion into competent hosts, he was also able to demonstrate a prolonged secretory activity in the ganglion. He demonstrated further that half a ganglion was

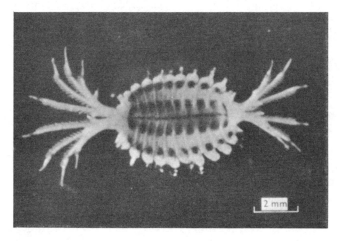

Fig. 8. A heteromorphic head produced at the posterior wound surface of an experimental animal exposed to 1.25×10^{-3} M colchicine for 6 days and then removed to IO. Both cephalic regions are regenerating simultaneously and are considered 'normal' (negatively phototactic).

incompetent while two halves achieved the same result as a whole normal ganglion; this demonstrates the presence of a threshold for activity of the ganglion. In sabellids, therefore, it may be possible that colchicine is reducing the rate of release of a neurohormone so that the time scale of regeneration is extended proportionately. This allows the essential tissue interactions to produce a normal regenerate when returned to IO before 5 days, and allows a build-up of subthreshold concentrations in tissues at the posterior wound region to produce a heteromorphic head when returned to IO after 5 days.

A peculiarity occurred if the animals were cultured in colchicine for periods greater than 5 days. In 10–15 % of the cases (6 out of 56 worms cultured), heteromorphic heads were produced at the posterior cut surface as well as normally at the anterior wound region upon return of the animals to IO (Fig. 8).

Thus two regenerating cephalic regions were produced simultaneously. The heads formed at a normal rate and were 'functional', i.e. negatively phototactic. Both cephalic regions induced parapodial inversion, characteristic of that type of regeneration. Reamputation produced two worms which regenerated at a normal rate and according to the *original* polarity of the animal, i.e. that cut surface which would normally support cephalic regeneration did so and vice versa. Thus two worms were obtained, one 'normal' and another with two heads. It is quite obvious that the colchicine affects the polarity for a brief, but critical, time and that the effects are not permanent. Furthermore, the ability to produce a cephalic region is not limited to the anterior cut surface.

Comparing this observation with the previously described tissue interactions, the development of bipolar worms appears plausible since both wound surfaces possess a severed nerve-cord, gut and body wall, all in their proper spatial relationship. However, during normal regeneration, polarity is expressed in an asymmetrical pattern which produces an anterior head and posterior pygidium. The administering of colchicine may block neurosecretion, and thus permit a different interaction to occur at the posterior wound region. Flickinger & Coward (1962) have obtained similar heteropolar regenerates in the platyhelminthes, but their worms were exposed to Colcemide for only brief periods, generally a day, and the animals were cultured *in* an agar slant.

It is of interest that these heteromorphic worms no longer possess a growth region. Normally, elongation of the worm is accomplished by cell division and growth in a region just anterior to the pygidium (Dales, 1963). The formation of a heteromorphic head may necessitate the omission of a growth area and therefore commit the worm to a static size. What physiological effects this might have are unknown at the present time.

SUMMARY

1. Gut and nerve-cord interaction during regeneration in the marine polychaetes *Sabella* and *Branchiomma* was examined. Surgical manipulation of various organ systems revealed a mutually dependent tissue interaction of gut, nerve-cord, and body wall in order for normal regeneration to occur.

2. Exposure of *Sabella* and *Branchiomma* to $1·25 \times 10^{-3}$ M colchicine prevents regeneration for a period of time which is directly proportional to the time cultured in colchicine. Removal from colchicine to normal sea water (IO) initiates regeneration without any evidence of sublethal cytolysis.

3. Bipolar heteromorphic heads are observed in 10–15 % of the organisms cultured in colchicine for more than 5 days. Reamputation after returning the worms to IO reveals that the effect is transient, the worms regenerating according to their original polarity.

4. The possible interaction of gut, nerve-cord, and body wall with a neurohormone is discussed. It is concluded that if regeneration in sabellids is under

neurohormonal control, it is possible that colchicine may retard the rate of release of the hormone and therefore permit the regenerating organ systems at the posterior wound region to produce a heteromorphic head.

The authors wish to thank Mr Ryland Loos for the technical diagrams, Mr Robert Speck for assistance with the photographs, and Mr Jörg Schulz for aid with German references.

This work was supported by Grant 20–216 from the Research Foundation of the State of New York and by a program-project grant from the National Institute of General Medical Science (PO 1 GM 14891–02) administered by Dr Robert D. Allen.

REFERENCES

Avel, M. (1932). Sur une expérience permettant d'obtenir la régénération de la tête en l'absence certaine de la chaîne nerveuse ventrale ancienne chez les lombriciens. *C. r. hebd. Séanc. Acad. Sci., Paris* **194**, 2166–8.

Avel, M. (1947). Les facteurs de la régénération chez les Annelides. *Rev. suisse Zool.* **54**, 219–35.

Bailey, P. L. (1930). The influence of the nervous system in the regeneration of *Eisenia foetida. J. exp. Zool.* **57**, 473–509.

BAILEY, P. L. (1939). Anterior regeneration in the earthworm, *Eisenia*, in the certain absence of central nervous tissue at the wound region. *J. exp. Zool.* **80**, 287–98.

BERRILL, N. J. (1931). Regeneration in *Sabella pavonina* (Sav.) and other sabellid worms. *J. exp. Zool.* **58**, 495–523.

BERRILL, N. J. (1952). Regeneration and budding in worms. *Biol. Rev.* **27**, 401–38.

BERRILL, N. J. (1961). *Growth, Development, and Pattern.* San Francisco: W. H. Freeman.

BERRILL, N. J. & MEES, P. (1936). Reorganization and regeneration in *Sabella*. I. Nature of gradient, summation, and posterior reorganization. *J. exp. Zool.* **73**, 67–83.

CLARK, M. E. & CLARK, R. B. (1962). Growth and regeneration in *Nephtys*. *Zool. Jb. Physiol.* **70**, 24–90.

CLARK, R. B. & BONNEY, D. G. (1960). Influence of the supraesophageal ganglion on posterior regeneration in *Nereis diversicolor*. *J. Embryol. exp. Morph.* **8**, 112–18.

CLARK, R. B. & CLARK, M. E. (1959). Role of the supraesophageal ganglion during the early stages of caudal regeneration in some errant polychaetes. *Nature, Lond.* **183**, 1834–5.

CROWELL, P. S. (1937). Factors affecting regeneration in the earthworm. *J. exp. Zool.* **76**, 1–34.

DALES, R. P. (1963). *Annelids.* London: Hutchinson.

FAULKNER, G. H. (1932). The histology of posterior regeneration in the polychaete *Chaetopterus variopedatus*. *J. Morph.* **53**, 23–58.

FLICKINGER, R. A. & COWARD, S. J. (1962). The induction of cephalic differentiation in regenerating *Dugesia dorotocephala* in the presence of the normal head and in unwounded tails. *Devl Biol.* **5**, 179–204.

GOLDFARB, A. J. (1914). Regeneration in the annelid worm *Amphinoma pacifica* after removal of the central nervous system. *Papers Tortugas Lab.* **6**, 97–102.

GOLDING, D. W. (1967a). Endocrinology, regeneration and maturation in *Nereis*. *Biol. Bull. mar. biol. Lab., Wood's Hole* **133**, 567–77.

GOLDING, D. W. (1967b). Neurosecretion and regeneration in *Nereis*. I. Regeneration and the role of the supraesophageal ganglion. *Gen. Comp. Endocrin.* **8**, 348–55.

GOLDING, D. W. (1967c). Neurosecretion and regeneration in *Nereis*. II. The prolonged secretory activity of the supraesophageal ganglion. *Gen. Comp. Endocrin.* **8**, 356–67.

GROSS, F. & HUXLEY, J. S. (1935). Regeneration and reorganization in *Sabella*. *Wilhelm Roux Arch. EntwMech. Org.* **133**, 582–620.

HERLANT-MEEWIS, H. (1964). Regeneration in annelids. In *Advances in Morphogenesis*, vol. IV (ed. Abercrombie and Brachet), pp. 155–215. New York: Academic Press.

HUNT, H. R. (1919). Regenerative phenomena following the removal of the digestive tube and the nerve cord of earthworms. *Bull. Mus. comp. Zool. Harv.* **62**, 571–81.

HUXLEY, J. S. & GROSS, F. (1934). Regeneration und 'Organisatorwirkung' bei *Sabella*. *Naturwissenschaften* **22**, 456–8.

KIORTSIS, V. & MORAITOU, M. (1965). Factors of regeneration in *Spirographis spallanzanii*. In *Regeneration in Animals and Related Problems* (ed. Kiortsis and Trampusch), pp. 250–61. Amsterdam: North Holland Publishing Co.

KROEBER, J. (1900). An experimental demonstration of the regeneration of the pharynx of *Allolobophora* from the endoderm. *Biol. Bull. mar. biol. Lab., Wood's Hole* **2**, 105–10.

KROPP, B. (1933). Brain transplantation in regenerating earthworms. *J. exp. Zool.* **65**, 107–29.

LEHMANN, F. E. (1957). Die Schwanzregeneration der Xenopuslarve unter dem Einfluss phasenspezifischer Hemmstoffe. *Rev. suisse Zool.* **64**, 533–46.

LEHMANN, F. E. (1965). Biochemical problems of regeneration. In *Regeneration in Animals and Related Problems* (ed. Kiortsis and Trampusch), pp. 56–67. Amsterdam: North Holland Publishing Co.

LENDER, TH. (1962). Factors in morphogenesis of regenerating fresh-water planaria. In *Advances in Morphogenesis*, vol. II (ed. Abercrombie and Brachet), pp. 305–31. New York: Academic Press.

LIEBMANN, E. (1942a). The role of the chloragogue in regeneration of *Eisenia foetida* (Sav.). *J. Morph.* **70**, 151–87.

LIEBMANN, E. (1942b). The coelomocytes of Lumbricidae. *J. Morph.* **71**, 221–49.

MORGAN, T. H. (1902). Experimental studies of the internal factors of regeneration in the earthworm. *Wilhelm Roux Arch. EntwMech. Org.* **14**, 562–91.

73

NICOL, E. A. T. (1930). The feeding mechanism, formation of the tube, and physiology of digestion in *Sabella pavonina*. *Trans. R. Soc. Edinb.* **56**, 537–98.

OKADA, Y. K. (1938). An internal factor controlling posterior regeneration in syllid polychaetes. *J. Mar. biol. Ass. U.K.* **23**, 75–8.

OKADA, Y. K. & KAWAKAMI, I. (1943). Transplantation experiments in the earthworm. *Eisenia foetida* (Savigny), with special remarks on the inductive effect of the nerve and on the differentiation of the body wall. *Tokyo Univ. Fac. Sci. J.* series 4, **6**(1), 25–96.

OKADA, Y. K. & TOZAWA, H. (1944). Supplementary experiments of transplantation in the earthworm: the induction of tail by the transplanted nerve cord. *Tokyo Univ. Fac. Sci. J.* series 4, **6**(5), 635–47.

SAYLES, L. P. (1932). External features of regeneration in *Clymenella torquata*. *J. exp. Zool.* **62**, 237–57.

SCULLY, U. (1964). Factors influencing the secretion of regeneration promoting hormone in *Nereis diversicolor*. *Gen. Comp. Endocrin.* **4**, 91–8.

STEPHAN-DUBOIS, F. (1958). Le rôle des leucocytes dans la régénération caudale de *Nereis diversicolor*. *Archs Anat. microsc. Morph. exp.* **47**, 605–52.

TAYLOR, E. (1965). Inhibition of mitosis. I. Kinetics of inhibition and the binding of ^{3}H-colchicine. *J. Cell Biol.* **25**, 145–60.

Later Stages of Regeneration in the Polychaete, Nephtys [1]

MARY E. CLARK

The early stages of caudal regeneration in the polychaete *Nephtys* have been described in earlier papers covering the period of wound healing and blastema formation (Clark and Clark, '62; Clark, '65).

The present report is concerned with regeneration subsequent to this stage and deals both with the proliferation of new segments, and with changes related to regeneration which occur in the animal as a whole. Since differentiation of individual tissues is more spaced out in regenerating segments than in normally growing animals, it is possible to observe the relative timing of the various morphological changes involved in formation of a new segment during regeneration. This aspect of segment proliferation is stressed here, in order to determine possible key points in the growth pattern.

MATERIAL AND METHODS

The animal used in this study was *Nephtys hombergi*. Specimens were collected intertidally from the Severn Channel. Uniformly small animals were chosen, although in *Nephtys* size is not closely correlated with the degree of sexual maturity. Since the gametes are not released into the coelom until just before spawning, sampling of coelomic fluid to determine maturity of individuals is not possible. The animals used here were collected between July and February and had either immature or, at most, only half-developed gonads. Amputation of posterior segments was obtained by autotomy without prior anesthesia. Worms were kept individually without food at 12 to 15° C, and sea water was changed every other day. In some instances the original "tail" was fixed at the moment of amputation. Individuals were allowed to regenerate from 5 to 21 days and were then fixed and stained by the methods used previously (Clark and Clark, '62).

DESCRIPTIVE FINDINGS

1. Early stages of new tissue formation: the pygidium

The rate of wound healing and the size of the blastema that is formed vary with the size of the individual, so that no absolute times for these events can be given. In our previous studies (Clark and Clark, '62) we were dealing primarily with large animals in which the rate of healing is slower and subsequent stages in regeneration are temporarily delayed. The size of the blastema is a function of the size of the wound, being larger in large animals

[1] This work was supported by funds from the Science Research Council of Great Britain.

or in those where accidental eversion of the gut occurs at the time of amputation. A large blastema may, in fact, hinder regeneration, partly by temporarily occluding the anus, and by preventing the free extension of the intestinal stump into the growing regenerate.

Previous observations of five-day regenerates in small worms similar to those used in this study show that healing is more rapid and the blastema is smaller than in larger worms (Clark, '65). The detailed processes of regeneration in small worms are the same as those previously described (Clark and Clark, '62) except that they occur more rapidly. Thus, not only is healing generally complete by about the fifth day, but also the ventral bud of a new anal cirrus is usually formed, and there are many large basophilic, mitotically active cells in the posterior epidermis (Clark, '65). In advanced five-day regenerates it is also possible to distinguish, at the level

of the anal cirrus bud and immediately ventral to the new anus, a zonation of the mesenchymal blastema (fig. 1). In the midline, the most posterior blastema cells are starting to differentiate a new anal muscle. These fibroblasts have irregularly shaped nuclei and muscle fibers are already differentiated in the cytoplasm. Blastema cells lying anterior to them, and also those lying laterally against the basement membrane of the regenerating epidermis, are highly enlarged, with large nuclei containing large nucleoli and basophilic cytoplasm. These cells are mitotically active and give rise to the mesodermal structures of the regenerate. More anterior still is a zone of small, fibrous mesenchyme cells having the typical appearance found in the early blastema (Clark and Clark, '62). These cells form a complete barrier between the gut and the body wall, and ventrally they completely cover the posterior face of the terminal pseudo-septum.

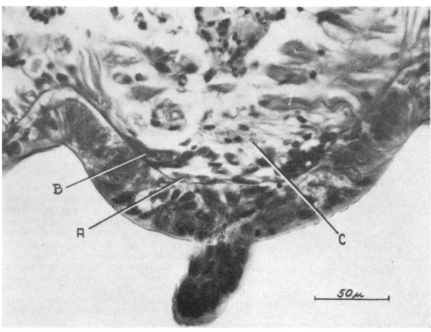

Fig. 1 Photomicrograph of frontal section of 5-day regenerate showing zonation of mesenchymal blastema. A, anal muscle; B, cells of regenerate; C, anterior region.

They function in healing by preventing loss of coelomic fluid and by drawing the edges of the wound together. They do not participate further in regeneration, but remain for a time marking the position between the old and new segments (fig. 13). Later they gradually disperse, reverting to coelomocytes (fig. 14). A regenerated basement membrane is now clearly visible in most areas between epidermis and mesoderm, and a new pair of enlarged terminal blood vessels is regenerated around the gut.

The pygidium and its associated structures, the anal muscle and anal cirrus, are the first structures to be regenerated, and the pygidium and anal muscle begin to show signs of differentiation at a time when no cellular differentiation is occurring in the young segments. As noted above the cells of the new sphincter muscle differentiate distinct muscle fibers within one week. There is thus a clear line of demarcation between the pygidial region and the zone of segment proliferation just anterior to it. With the regeneration of the anal sphincter muscle a normally functioning anus is restored. Further development of the anal muscle during the following weeks of regeneration is restricted to an enlargement in its antero-posterior dimension.

During the second week of regeneration, the epidermal cells of the lateral and dorsal parts of the pygidium lose their cytoplasmic basophilia, fibers appear in the cytoplasm and the nuclei become smaller and less regular than those in the adjacent epidermis of the segment-proliferating region. At this time the epidermis dorsal to the anus shows the distinctive pleated pattern characteristic of the intact animal (see Clark and Clark, '62). On the other hand, mitotic activity remains high for 2 to 3 weeks in both the ventral cushion of the pygidium, which is continuous with the regenerating nerve cord, and the anal cirrus which continues to grow rapidly (see table 2).

It is probable that fibers of the ventral nerve cord enter the bud of the anal cirrus as soon as it begins to form. They are readily seen by the second week of regeneration, when paired fiber bundles are visible, growing from the neuropile into the ventral flap of the pygidium. Their position may be seen in the 20-day regenerate shown in figure 6, I to L.

Also making an early appearance in the regenerating pygidium is a special cell type which normally occurs in the basal regions of both the epidermal and endodermal epithelia and also in the ventral nerve cord and along the segmental nerves (see Clark and Clark, '62, where they were described in the nerve cord as a special type of glial cell). These cells are generally small, elongate, and lie parallel to the basement membrane. They have small, darkly staining irregularly shaped nuclei and are readily identified by the numerous fine paraldehyde-fuchsin positive granules dispersed uniformly through the cytoplasm (fig. 2). In addition to their possible nutritive role in the nerve cord, these cells exhibit strong phagocytic properties and possibly are able to move about among the other cells. They phagocytoze pycnotic cells and remove debris resulting from wounds. In all their properties they closely resemble the free coelomocytes, the latter differing from them only in their location, in their greater irregularity of outline, and in the presence of fine fibers in the peripheral parts of their pseudopodia (see Clark and Clark, '62). Cells of this type are already present in the epidermis which grows over the wound where they multiply by mitotic division both in the pygidium and in the epidermis of young segments.

The regeneration of specialized ciliated epidermal cells in the pygidium also precedes their appearance on the parapodia of regenerated segments where they normally function in producing respiratory currents. These cells are characterized by an extremely large nucleus, containing a very large nucleolus. The cluster of cilia leaves the body via a modified gap in the cuticle (fig. 3). A pair of such cells may begin to differentiate, one on each side of the regenerating pygidium by the ninth day, and fully developed ciliated cells usually appear by the twelfth day of regeneration. A few begin to differentiate on the regenerated parapodia of the first segment or two, but only after two weeks.

The pygidial blood sinus appears very early, as soon as the first sign of pygidial regeneration. The terminal ends of the sub-intestinal vessel and the longitudinal

77

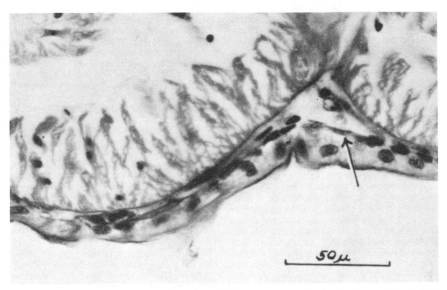

Fig. 2 Photomicrograph of frontal section of 5-day regenerate showing basal, paraldehyde-fuchsin positive cells (arrow) migrating with other epidermal cells towards wound.

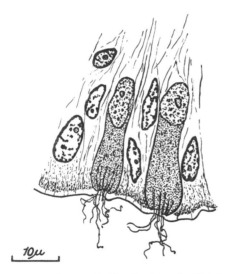

Fig. 3 Camera lucida drawing of ciliated sensory cells in the pygidium of a 14-day regenerate.

vessel of the blood plexus in the ventral wall of the gut enlarge and become confluent at their point of contact with the regenerated epidermal basement membrane. There they grow downward between the anal muscle and the pygidial epidermis to form the pygidial sinus, which enlarges later to the base of the anal cirrus ventral to the gut (see fig. 6). The sinus always contains blood of an "older" appearance than the finely granular blood which fills the youngest segmental vessels just anterior (see Clark and Clark, '62), making yet another contrast between the differentiated pygidium and the undifferentiated young segments just anterior.

Thus, while at five days only a small, undifferentiated pygidial rudiment has regenerated, after one and a half weeks of regeneration, at a time when several young segments have been proliferated which are still in a visibly undifferentiated state, the pygidium, although it is the most caudal organ of the regenerate, is already nearly fully differentiated. Only the anal cirrus and the ventral cushion, which produces epidermal cells for the young segments, contain a large proportion of undifferentiated cells.

The one difference between a regenerated pygidium and that of an intact animal which is still evident after three full weeks of regeneration is its unusual proportions. The regenerated pygidium is much smaller and, especially, much narrower. The lateral parts of the pygidium, where mitoses are normally rare in intact worms, have a much reduced rate of mitotic activity in the regenerate after it starts to differentiate, and its further growth is slow. The rapidly growing ventral cushion of the pygidium and the base of the anal cirrus, however, soon approximate to their full size, and hence look disproportionately large in the regenerate.

2. Segment proliferation and continuity of non-segmental structures

a. *Relationship between mitosis, segmentation and cytodifferentiation*

Segment formation usually begins as soon as wound healing is complete, but according to the size of the wound and disposition of the gut, the first appearance of new segments may vary in both time and arrangement. If the severed end of the intestine remains *in situ* and the wound itself is small, healing is effected quickly, the posterior continuity of the epidermal and intestinal basement membranes is soon re-established, and early signs of segment proliferation may appear at the end of the first week of regeneration. Since the volume of the regenerate at this time is still very small, only one or two segments are formed initially and more are added as soon as mitotic divisions have produced enough new cells for the purpose.

On the other hand, if the wound was enlarged by the extrusion of the gut, healing is much slower, and more than a week may be required for the epidermis to grow posteriorly and cover over the exposed gut. Until this occurs and the two basement membranes are rejoined, no segment formation occurs, the exposed zone between gut and epidermis being still filled with attached coelomocytes. As the wound closes, the coelomocytes begin to disperse, and when healing is complete a cylinder

of new epidermis is left surrounding the gut, in which several segments have formed almost simultaneously. By this means, the early retardation in segment formation is compensated, so that by two weeks animals with both types of wound healing show approximately equivalent segment regeneration.

Regenerated segments are formed in exactly the same manner as are new segments during normal growth of the worm (Clark and Clark, '62) except that the rate of segment formation is much faster during regeneration, so that the difference in size and development of successive segments is much less than in the normal growing tip. This difference results from the much higher rate of cell proliferation in the early regenerate, where the balance between cell production and cell growth swings in favor of the former. The nuclear to cytoplasmic volume ratio is very much increased throughout the regenerate and it is only when the total number of mitoses in the regenerate decreases (fig. 4) that the cytoplasmic growth of individual cells begins to restore the balance and the older segments of the regenerate enlarge rapidly. At this time the rate of segment proliferation shows a decrease (fig. 5).

The visible differentiation of individual cells is slightly retarded during regeneration when compared to cells in the same position relative to their distance from the pygidium in the normal growing tip. Likewise, the first appearance of morphological structures may be delayed slightly during this early stage of regeneration when segment proliferation is most rapid (table 1). This may not reflect so much a decreased rate of cell differentiation but rather an increased rate of cell proliferation posteriorly, so that new segments are formed before the old ones have time to attain an advanced level of differentiation. Not all tissues differentiate at the same rate, as will become apparent below (see table 1).

b. *Mesodermal structures*

(1.) *Role of the blood-vascular system.* The first sign of the proliferation of a new segment is the formation on the anterior face of the anal sphincter muscle of a

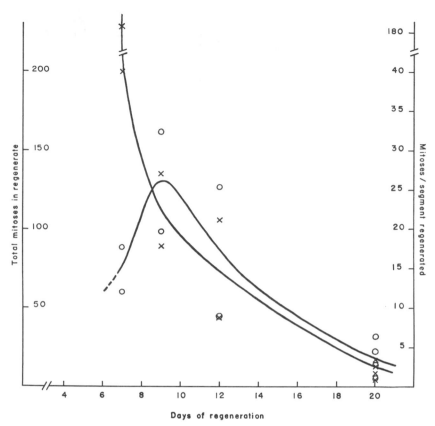

Fig. 4   Total number of mitoses (O) and mitoses per segment regenerated (X) as a function of time.

new segmental blood vessel and its associated fibroblasts which span the coelom from gut to body wall. The development of these structures may be followed in the serial diagrams of the posterior part of the 20-day regenerate shown in figure 6. The ninth regenerated segment is clearly separated off from the growing point, and a space, the coelomic cavity of the tenth segmental blood vessel is only beginning to be formed on one side (fig. 6E,F).

The formation in rapid succession of several new segmental blood vessels results from a marked response from the blood vascular system, and throughout the first

two weeks of regeneration the blood vessels of the regenerate are disproportionately large. The healed ends of the longitudinal blood vessels above and below the gut are very much distended during the first week of regeneration, and a new pair of enlarged vessels encircling the gut forms, due to lateral expansion of the terminal region of the vascular plexus which lies in the gut wall (Clark and Clark, '62). During the second week of regeneration, at the time of most rapid segment proliferation, this entire caudal vascular complex swells, and both the dorsal and ventral surfaces of the gut become covered by a distended blood

80

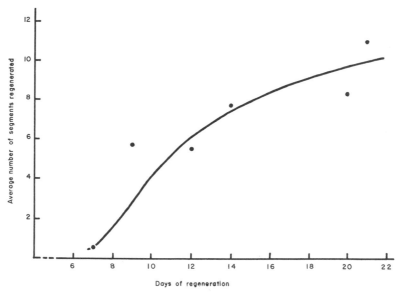

Fig. 5 Number of segments regenerated as a function of time.

TABLE 1

Relative rates of tissue differentiation during regeneration expressed as the distance in segments anterior to the pygidium in which each tissue first appears

Age of regenerate	Number of segments re-generated[1]	Body-wall muscle fibers		Chaetal sac		Aciculum		Chaetae	
days									
(intact)	—	(1)[2]		(2)		(3)		(4)	
7	0	—	(1)	—	(1)	—	(2)	—	(2)
7	0.5	—	(1)	—	(1)	—	(2)	—	(3)
7	0.5	—	(1)	—	(1)	—	(2)	—	(3)
7	1.5	—	(2)	—	(2)	—	(3)	—	(4)
9	5.5	4.5		3.5		5.5		—	
9	6.0	4		4		6		6	
12	5.0	3		4		5		—	
12	6.0	4		3		4		5	
14	7.5	4.5	(1)	3	(2)	4.5	(2)	6.5	(3)
14	8.0	5	(1)	3	(2)	4	(2)	5	(3)
20	6.5	2.5		3.5		3.5		4.5	
20	8	2		3		4		5	
20	9	2		3		4		5	
20	9.5	3.5		3.5		5.5		6.5	
21	10.0	2	(1)	3	(2)	4	(3)	5	(3)
21	12.0	4	(2)	4	(2)	4	(3)	5	(4)

[1] Half-segments indicate incomplete separation of segmental blood vessel and coelom.
[2] Figures in brackets are from original amputated tails.

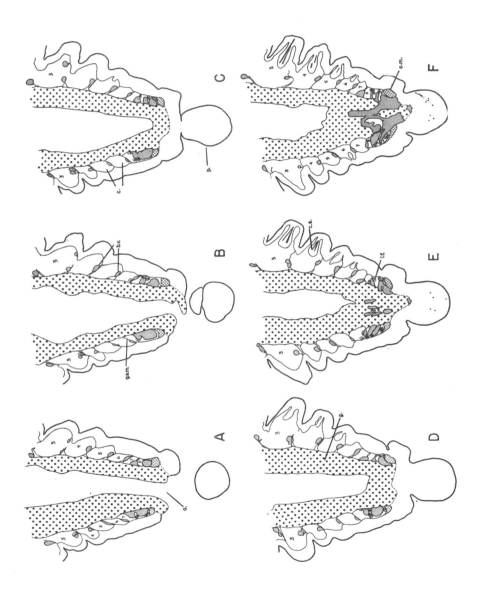

82

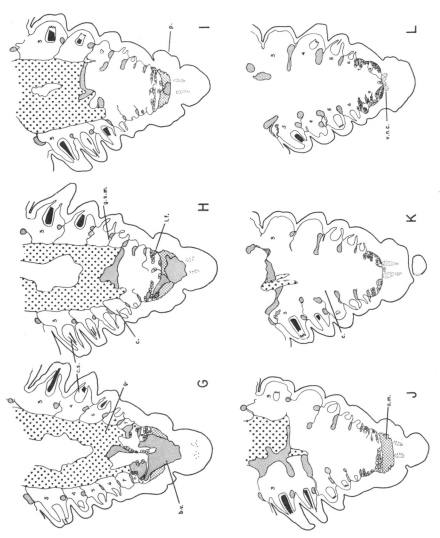

Fig. 6 A-L Semi-diagrammatic interpretation of camera lucida drawings of dorsal to ventral series of frontal sections of posterior region of 20-day regenerate. a, anus; a.m., anal muscle; b.v., blood vessel; c., coelom; c.s., chaetal sac; g, gut; g.s.m., gut suspensory muscle; l.f., large fibroblasts; p, pygidium; v.n.c., ventral nerve cord.

sinus. Laterally this gradually becomes divided into distinct segmental vessels (see below). The dorsal vessel often extends half-way around the gut (fig. 7) while the swollen sub-intestinal vessel may become bifurcated into two parallel channels under the elongated gut (fig. 6F), resembling the condition often existing in young segments of intact animals (Clark and Clark, '62).

New segmental blood vessels form by the splitting of the terminal circum-intestinal vessels into an anterior and posterior vessel. At first the two vessels are separated only laterally, perhaps being pinched off by the inward growth of large fibro-

blasts. Gradually the cleavage extends, first ventrally and then dorsally, around the gut towards the midline. The space thus formed between this new blood vessel and the terminal one from which it has just split off is the new coelom of the most posterior segment (fig. 6F). As the new segment grows, the vessels grow laterally also and separate away from the side of the gut wall, until eventually they retain contact with the latter only in the dorsal and ventral midlines. Although the new vessels are no larger than the segmental blood vessels of the stump, their disproportionate size in the tiny new segment results in a close contact with the lateral

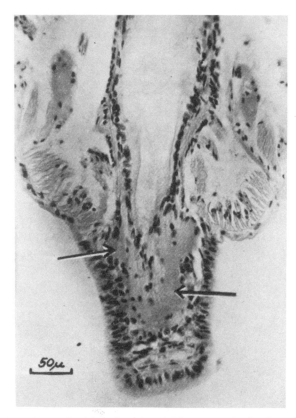

Fig. 7 Photomicrograph of frontal section of 9-day regenerate showing enlarged dorsal blood vessel (arrows) overlying gut.

body wall of the new segment to which they are attached by delicate fibers. These fibers later form the mesentery-like membrane by which the nephridial vessel, which is derived from the vertical portion of the segmental loop, is suspended from the posterior wall of the parapodium. In other regions of the parapodium, branches from the segmental vessel actually fuse with the epidermal basement membrane (Clark and Clark, '62).

By the end of the third week of regeneration, when the rate of segment proliferation has slowed, most of the regenerated segmental blood vessels are reduced in absolute size, yet despite further growth of the segments they are still disproportionately large compared to those in the anterior intact segments. The two terminal pairs of segmental blood vessels may also be less enlarged than earlier. The sub-intestinal blood vessel in the anterior part of the regenerate region and also in the stump often becomes much reduced. The dorsal vessel on the other hand remains large throughout the regenerate region, right through the twentieth day, and is sometimes distended anteriorly in several segments of the stump.

The nephridial vessels of the last stump segment partially collapse during the first few days of regeneration (Clark and Clark, '62), but by the time segment proliferation begins they are enlarged beyond their normal size. During the second week they grow posteriorly into the first regenerated segment, at the same time as the longitudinal paraneural vessels regenerate into it.

Like the other longitudinal vessels, the paraneural vessels are healed within a few hours of amputation, and also have swollen ends containing numerous blood cells (Clark and Clark, '62). But unlike the longitudinal vessels associated with the gut, the paraneural vessels do not participate in the early stages of regeneration; their swollen tips shrink and sometimes collapse altogether around the enclosed blood cells. The first sign of regeneration of a paraneural vessel is the appearance at the healed tip of an elongate cord of large fibroblasts which grows into the new segments alongside the regenerating nerve cord and probably forms appropriate connections with the regenerating segmental

vessels at this time. Blood first appears in the lumina of these vessels relatively late.

The walls of newly regenerated blood vessels appear to be extremely fragile and susceptible to damage during fixation. The blood which is found in the new vessels also has a peculiar granular consistency after fixation; this is described as "new blood," and seems to be common to recently formed vessels in both intact (Clark and Clark, '62) and regenerating animals. It contrasts with the intensely and uniformly staining "old blood" in the vessels of the stump. There appears to be little mixing of blood between the old and regenerate regions until the end of the third week of regeneration when "old blood" may fill the larger vessels of the regenerate. Whether the blood which first fills the new vessels is produced locally or is in some way altered, perhaps by dilution, as it passes from stump to regenerate, is difficult to determine. It is likely that the protein content of "new blood" is comparatively low. These points will be discussed more fully below when changes in the nature of the blood cells in the dorsal vessel of the stump are considered.

(2.) *Muscle regeneration.* It is apparent in figure 6 that the most caudal part of the segmental regenerate, that is the region with the greatest number of segments, is between the ventral half of the gut and the dorsum of the nerve cord. The largest, most actively dividing fibroblasts of the regenerate are in this area. These cells appear to be derived from the middle or regeneration zone of the early blastema, described above, and they probably give rise to all the mesodermal structures of the regenerate: muscle, coelomic epithelium of various types, nephridia, and eventually, gonads. These cells also have phagocytic properties, and frequently have vacuoles containing partially digested remnants of damaged or pycnotic cells (fig. 8).

During the second week of regeneration, these large fibroblasts are numerous throughout all of the regenerated segments but are most frequent ventrally. They usually occur along the surfaces of blood vessels, especially the tip of the sub-intestinal blood vessel. By the end of the third week of regeneration, however, most of

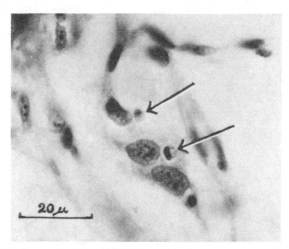

Fig. 8 Photomicrograph of large fibroblasts with phagocytozed material in vacuoles (arrows).

the mesodermal cells in the older part of the regenerate which are not already differentiating into muscle cells are nevertheless reduced in size, have small nuclei, and resemble ordinary coelomic epithelium. In the youngest, most posterior segments, large fibroblasts are still to be found, but the area of their distribution is now limited to the anterior face of the anal muscle just ventral to the gut, and to the surface of the sub-intestinal blood vessel in the last few segments (fig. 6G). In the growth zone of intact animals, only two or three cells of this type normally occur.

The first muscle formed in each new segment, identifiable by its association with the segmental blood vessel before any signs of cytological differentiation, is the gut suspensory muscle, the true septum in *Nephtys* (see Clark, '62; Clark and Clark, '62) (fig. 6F). The anlagen of these muscles are laid down ventro-laterally, but move dorsally as the segment grows (see fig 6). As with all the other muscles, they begin as undifferentiated fibroblasts, whose attachments to basement membranes establish the origin and insertion of the future muscle. Only later do muscle fibers appear in the cytoplasm.

The large fibroblasts which line the external surfaces of the coelom in the young segments form attachments between various structures and become the transverse, dorso-ventral and parapodial muscles. Due to the epidermal location of the nervous system in *Nephtys* all these muscles have at least one insertion on the epidermal basement membrane (the one exception being the muscle coats investing the gut which probably receive stomatogastric innervation) (Clark, '66).

Likewise the longitudinal muscles are regenerated from fibroblasts lying along the dorsal and ventral coelomic surfaces of the youngest segments. Because new segments are formed ventrally and later rotate dorsally by differential growth, there are always fewer segments dorsal to the gut then ventral to it (fig. 6), and for some time the anus in fact opens almost dorsally. For this reason the dorsal longitudinal muscle never spans as many segments in the regenerate as the ventral one, which appears on the ventral body wall during the second week of regeneration and continues to elongate at the same rate as the regenerate. From this time on, as soon as new fibroblasts become associated with either dorsal or ventral longitudinal muscles they begin to differentiate muscle fibers.

Continuity between the longitudinal muscles of the stump and the regenerate is established in the following way. Not all muscle fibers of the last segment in the stump are severed by amputation since the most peripheral fibers have their posterior insertions on the intact epidermal basement membrane in that segment (fig. 9). More medially, damaged fibers are either removed by coelomocytes or dedifferentiate to fibroblasts. The latter, together with coelomocytes, serve to connect the severed ends of the muscles to the blastema and to the basement membrane of the posterior epidermis as it grows over the wound (Clark and Clark, '62). Some of these cells contribute to the regeneration of the longitudinal muscles. The posterior end of the cell attaches to the basement membrane of the regenerate epidermis and the other inserts among the muscle fibers remaining in the stump. Since the number of fibroblasts contributing to the new longitudinal muscles is restricted by the small volume of the regenerate, only a small proportion of the muscle fibers in the stump region which were severed on amputation is replaced during the first weeks of regeneration. These are located just internal to the intact fibers in the stump, primarily in the median region of the muscle mass (alongside the nerve cord in the case of the ventral muscles) (fig. 9).

After three weeks of regeneration, the removal of damaged fibers is virtually complete and continuity between the longitudinal muscles of the regenerate and the stump is established, although the fiber density in the muscles of the last segment or two of the stump is still reduced (fig. 10). That the new muscle cells are still in the process of active growth, despite the differentiation of muscle fibers within them, is attested to by their relatively high ratio of sarcoplasmic to fiber volume, and by their larger nuclei and nucleoli compared to the fully differentiated cells of the stump.

(3.) *Other segmental structures.* After three weeks of regeneration other segmental structures are beginning to differentiate in the more anterior segments. The rudiments of nephridia (solenocytes, a very small coelomostome, and tiny nephridial duct) may be found in the anterior two or three segments. A complex of gonadial blood vessels may sometimes be distinguished on each side in the older segments, but no signs of gametogenesis are found. Likewise, newly differentiated segmental pyriform cells (see Clark and Clark, '62) in the first five or six regenerated segments are beginning to show signs of secretion by the end of the third week, although the cells are still very small. The secretion does not fill the entire cell, but only occurs in the periphery.

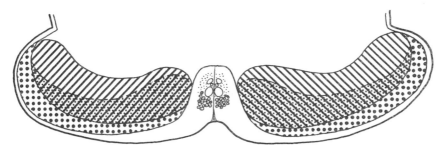

```
[····] Zone of intact fibers
[\\\\] Zone of severed fibers
[::::] Zone of regenerated fibers
```

Fig. 9 Schematic diagram of cross-section of ventral longitudinal muscle showing how fibers in terminal stump segment are affected by amputation and regeneration.

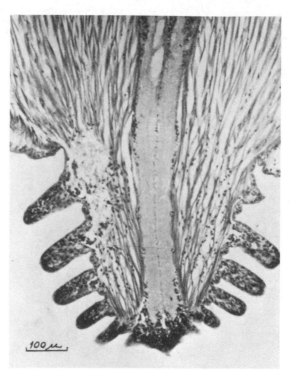

Fig. 10 Photomicrograph of frontal section of 21-day regenerate at level of ventral longitudinal muscles, showing reduced number of fibers in stump segments.

c. *Role of the intestine*

Whereas the blood vessels surrounding the intestine play a major part during the formation of new segments, the epithelium itself shows little activity during regeneration. Aside from a brief outburst of mitoses near the severed end of the gut during the first few days of regeneration (Clark and Clark, '62), there are very few mitoses in the endoderm compared with those in the epidermis and mesoderm during regeneration (table 2).

Very few gut cells are damaged at the time of amputation, and with the rejoining of the endodermal and epidermal basement membranes and restoration of the anus, healing of the gut is complete. As a result, there is no zone of large, dedifferentiated, basophilic and mitotically active cells in the gut. The small number of damaged cells is soon phagocytozed by the basally situated granular, paraldehydefuchsin positive cells described above.

During the posterior growth of the regenerate, the gut plays a nearly passive role. As the new segments are regenerated posteriorly, the gut in the last few segments of the stump is stretched longitudinally displacing the gut suspensory muscles in this region slightly posteriorly. Although the total diameter of the gut remains almost the same, the thickness of its epithelium in the last one or two segments of the stump and in the regenerate is considerably thinner than it is further anteriorly, the cross-sectional area of the lumen becoming consequently larger. Elasticity of the gut tissue is essential in an animal which

TABLE 2
Distribution of mitoses in the tissues of the regenerate

Age of regenerate	Number of segments re-generated [1]	Pyg.	VNC	Epid.	Gut	Meso.	Total	Mitoses/ segment regenerated
days								
7	0.5	20	22	30	11	5	88	176.0
7	1.5	24	8	14	5	9	60	40.0
9	5.5	20	19	32	2	25	98	17.8
9	6.0	11	32	77	4	38	162	27.0
12	5.0	10	10	13	1	10	44	8.8
12	6.0	18	25	48	1	35	127	21.2
20	6.5	1	1	2	0	1	5	0.8
20	8.0	2	0	6	5	0	13	1.6
20	9.0	6	2	6	5	3	22	2.5
20	9.5	13	4	10	3	2	32	3.4

[1] See footnote to table 1.

readily withstands a stretching in body length of 50%. Even though there may be up to twelve segments in the regenerate, its total length after three weeks is seldom much more than the length of one and a half or two segments in the stump, a distance requiring very few additional cells to permit stretching over the whole length of the regenerate. It appears that neither amputation nor stretching of the gut is an immediate stimulus to its growth, which may in fact be deferred until the arrival of the next wave of cell division in the normal intestinal cycle of secretion and replacement (Clark and Clark, '62).

d. *Regeneration of epidermal structures*

After the epidermis has completed its medial migration and covered the old wound it changes its character. The cells enlarge, become basophilic and are mitotically active; thus the whole wound epidermis resembles the small epidermal growth zone on the median postero-ventral surface of the normal pygidium (Clark and Clark, '62). These cells, however, retain their ability to secrete new cuticle, which begins to form as soon as healing occurs. At the time of its medial migration over the wound, the epidermis carries along its usual complement of basally situated, granular, paraldehyde-fuchsin positive cells (fig. 2). Such cells can always be found in regenerating epidermis and are mitotically active as well as phagocytic.

During the second week of regeneration a wave of vacuolization passes through the epithelium of the new epidermis. The cell nuclei are pushed peripherally to the cuticular side of the epidermis by the formation of large vacuoles located basally. The granular basal cells, however, are not affected and remain in their usual position. This vacuolization begins antero-dorsally and proceeds postero-ventrally, affecting all the new segments in turn during the intermediate stages of their differentiation. The tips of the parapodia, the chaetal sacs and the nerve cord are not involved in this wave of vacuolization, and by the third week of regeneration the epidermis in the older segments of the regenerate again has a normal appearance. If the vacuolization itself is an artifact of fixation it clearly has some basis in a temporary change in the nature of the epidermis, the significance of which remains obscure at present.

As the segments grow, the epidermal cells lose their basophilia and develop the intracellular fibers characteristic of epidermis, but even after three weeks, the cells are not fully grown and still have a distinctly high nuclear to cytoplasmic volume ratio. Differentiation of specialized cells begins at the end of the second week, when ciliated epidermal cells are sometimes identifiable on the most anterior segments of the regenerate. The time of their first appearance varies considerably among individual animals, however, and always lags behind the differentiation of similar cells on the pygidium (see above). Likewise, the differentiation of mucus cells is variable. After three weeks there is usually no sign of these cells in the epidermis, but in one animal tiny mucus cells could be seen on the posterior epidermis of the first four pairs of parapodia.

In terms both of the actual number of cells proliferated and the rapidity of differentiation, the chaetal sacs are the most conspicuous epidermal structures formed in each segment and appear at a very early stage in the development of a new segment (table 1). The first visible sign of a new chaetal sac is the inward growth of several large, undifferentiated epidermal cells into the ventral part of the coelom of a young segment (fig. 6 J, segments 7, 8). Probably both the dorsal and ventral sacs appear at the same time although it is not easy at first to distinguish them as separate entities. As the segment grows, the chaetal sacs enlarge and each soon begins to secrete an aciculum. Shortly afterward chaetae also begin to appear (fig. 6I; table 1). Yet despite the early onset of differentiation in the chaetal sacs, indicated by their secretory activity, they are still rapidly growing even after three weeks of regeneration; mitoses are fairly frequent and the cells show little visible sign of the cytoplasmic differentiation characteristic of fully grown segments.

e. *Regeneration of the nerve cord*

Although the nerve cord is an epidermal structure, it deserves separate consideration as being the zone of most intense mitotic activity during early regeneration (table 2) and of closest cell packing throughout the entire period of regeneration. As previously described, the cut axons of the neuropile are quickly healed and show little sign of serious degeneration (Clark and Clark, '62) although the giant axons often regress anteriorly from the plane of section by half a segment and their tips sometimes become swollen and vacuolated.

The first stage of regeneration of the nerve cord around the end of the first week of regeneration is the proliferation of a large number of small, closely packed cells in the ventral midline just anterior to the bud of the anal cirrus. As the regenerate grows, this zone of closely packed cells becomes longer and broader, its lateral margins merging imperceptibly with the epidermis ventral to the young longitudinal muscles. Throughout the whole period under study most of the cells in this area remain closely crowded and retain a high nuclear to cytoplasmic volume ratio. There is consequently a sharp line of demarcation between the old and new ganglionic regions of the nerve cord (fig. 11). Pycnotic cells frequently occur in this zone and are phagocytozed by the small, granular, paraldehyde-fuchsin positive cells of the nerve cord.

Regenerating neuropile is readily observed during the second week, although a few fibers may be present much earlier but escape detection with the histological methods used. The whole of the regenerate neuropile, even at the end of three weeks is covered dorsally only by basement membrane, and lacks the dorsal layer of large glial cells present in all but the most posterior one or two segments in the intact animal (Clark and Clark, '62). Presumably glial cells will eventually migrate dorsally to this position. As mentioned above, the neuropile innervates the pygidium and anal cirrus, probably near the time they are formed. The first appearance of segmental nerves in these preparations was generally in the third segment anterior to the growing point, but again, it is likely that smaller nerves innervating even younger segments exist but were not detected.

From the time they are first formed the fine fibers of the regenerate neuropile have an apparently unbroken continuity with those of the stump, but the regeneration of the giant axons is much slower and is only just beginning during the third week of regeneration. At first the whole of the regenerative neuropile is composed of fine fibers, but after about two weeks, in the ventro-medial region of the more anterior part of the neuropile where the giant axons will eventually form, irregularities appear among the fibers and they begin to disappear. By three weeks the new giant tracts are clearly defined as paler regions of the neuropile, although they still contain quite a few fine fibers and a thin transverse boundary of fibers continues to separate the old and new tracts (fig. 12).

As in the epidermis, the granular, paraldehyde-fuchsin positive cells are the first

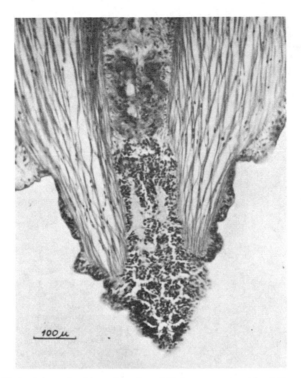

Fig. 11 Photomicrograph of frontal section of 21-day regenerate at level of ventral nerve cord showing morphology of old and new ganglionic regions.

cells to visibly differentiate. They appear during the second week of regeneration as faintly stained cells in the neuropile alongside the future giant axon tracts and accompanying the segmental nerves; a few also occur throughout the ganglionic layer of the regenerating nerve cord.

A small proportion of the remaining cells in the ganglionic layer of the anterior part of the regenerate begin to differentiate into unmistakable neurons by the end of three weeks, several pairs being found in those segments which have distinct segmental nerves. The cell bodies of the giant neurons are also readily identifiable in these segments. In no case were any signs of neurosecretory cells seen in the regenerate. The majority of cells in the ganglionic region remain closely packed together and give no indication of cytoplasmic differentiation, so that the nerve cord remains morphologically the least differentiated of all tissues in terms of the difference in appearance between a three-week regenerate and the stump. Yet the regeneration of a well developed neuropile suggests that these cells may already be able to function.

3. *Fate of the anterior zone of the blastema*

The coelomocytes which migrate to the wound appear to serve at least two functions in addition to possibly supplying cells for regeneration itself. By forming bridges between the gut and the body wall, they act in a purely mechanical way to bring the wound together and facilitate healing. In addition, their disposition at the wound is such that they form a complete screen

91

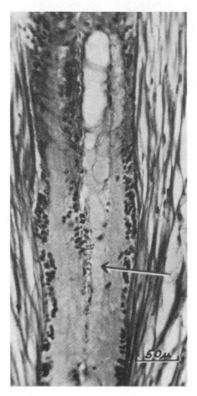

Fig. 12 Photomicrograph of frontal section of 21-day regenerate at level of ventral nerve cord showing early formation of giant axons (arrow) in regenerate.

chemical gradient responsible for the directional migration and attachment around the wound of the coelomocytes (Clark, '65), although the overall gradients in the coelomic fluid are not altered during the time the wound is open (Clark, '67). That a local alteration in the internal environment at the wound does occur is strongly suggested by the secretory activity of the segmental pyriform cells (see below). The role of the blastema in preventing loss of coelomic fluid and as an aid to healing is supported by two further observations: the number of coelomocytes contributing to the blastema is proportional to the size of the wound, and the anterior, non-regenerative part of the blastema gradually disperses after healing is complete.

The first blastema cells to free themselves from their attachments and revert to coelomocytes are those lying anterior to the terminal pseudoseptum of the stump; later the cells inserting between the dorso-ventral muscles of the last parapodium of the stump do likewise. The last cells to disengage themselves are those dorsal to the gut and those ventral to it and posterior to the terminal pseudoseptum which form a virtually complete transverse barrier between the old and new coelom (fig. 13). Even after three weeks, a few coelomocytes may still be found disengaging from the dorsal part of the gut at the level of amputation (fig. 14). On reverting to the coelomocyte condition the cells regain their characteristic cytoplasmic fibers and granules.

In addition to their purely mechanical functions, these cells may secrete some of the components of the coelomic fluid, but there is no direct evidence of this so far. They certainly have phagocytic properties during this period of attachment, and not only remove cellular debris from the coelom, but also phagocytoze bacteria whenever an infection occurs. In two cases of mild infection, bacteria were found in vacuoles of these blastema cells and also in the fixed phagocytes (Clark and Clark, '62) and in the granular, basal epidermal cells of the stump, but none were seen anywhere in the regenerate itself.

between the coelomic space in the stump and the exterior, hindering invasion by foreign organisms and preventing excessive loss of coelomic fluid before the wound is completely healed.

The importance of coelomic fluid in polychaete locomotion has been stressed (R. B. Clark, '64). Moreover, it contains considerable amounts of sugars, amino acids and other compounds which undoubtedly constitute a necessary *milieu* for the cells participating in regeneration (M. E. Clark, '64). It has been suggested that the temporary dilution of the coelomic fluid by sea water at the wound may provide a

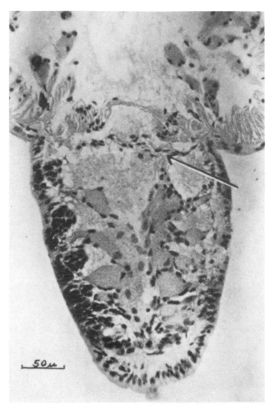

Fig. 13 Photomicrograph of frontal section of 9-day regenerate showing anterior blastema cells (arrow) still remaining attached at level of amputation.

4. *Changes in the animals as a whole*

a. *Segmental pyriform cells*

The extensive secretory response, following wounding, of the segmental pyriform cells in the last few segments of the stump has been described (Clark and Clark, '62). These large cells, lying in clusters suspended into the coelom from the base of each chaetal sac, appear to secrete at least some of the components of the coelomic fluid. Following loss of coelomic fluid at the wound they become so depleted of secretion as to lose their normal shape and staining properties. Similar activity can be elicited simply by withdrawal of coelomic fluid from intact animals (R. B. Clark, unpublished observations).

The rate of recovery of the pyriform cells after caudal amputation is variable and often differs even between cells in the same cluster, so that some of the cells are full of secretion and others have very little. The amount of secretion being released during the first week and a half of regeneration is usually greater than in intact animals, judging by the numerous secretion blebs over much of the cell surface. If the function assigned to them is correct, then not only do these cells have to compensate for any coelomic fluid lost from

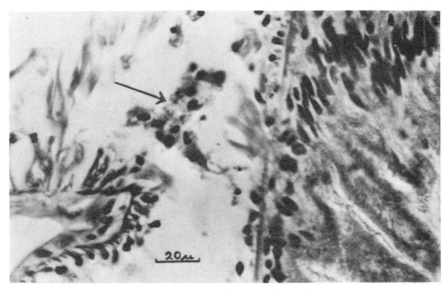

Fig. 14 Photomicrograph of frontal section of 21-day regenerate showing disengagement of anterior blastema cells (arrow) from wall of gut and body wall at level of amputation and their reversion to coelomocyte appearance.

the stump segments due to leakage from the open wound, but they are also the most probable source of the components of the coelomic fluid of the early regenerate before its own segmental pyriform cells begin to function. By two weeks, however, most of the segmental pyriform cells of the stump have regained a normal appearance, and during the third week the first differentiation of stainable material in the small pyriform cells of the regenerated segments suggests that, from this time on, the coelomic-fluid components of the regenerate are produced locally.

b. *Blood cells and the question of the formation of new blood*

The large numbers of free blood cells which appear at the tips of the severed blood vessels within a few hours of amputation (Clark and Clark, '62) do not appear to have an important role in the regeneration of the blood vessels in the regenerate. Whether they migrate posteriorly in response to the wound or are simply trapped passively due to the interrupted flow of blood, they remain at the tips of the longitudinal blood vessels for several days. They do not disperse immediately after the regeneration of new vessels, nor do they migrate in large numbers into the latter as might be expected if they were directly involved in the formation of new blood or vessels locally. The apparent lack of a vigorous circulation of blood between the vessels in the stump and in the regenerate would account for the absence of significant passive transfer of blood cells into the new vessels, and relatively few cells are found in them even after extensive regeneration. The cells which accumulate at the level of the wound gradually disperse during the second week of regeneration.

In both intact and regenerating animals, however, the blood cells in the dorsal longitudinal vessel are always more numerous than elsewhere, although distribution may be irregular along the length of the vessel. The majority of these cells lies along the ventral surface of the vessel, closely apposed to the gut wall, and they show some interesting changes during the course of

94

regeneration. In intact animals, the proportion of paraldehyde-fuchsin positive cells among the total blood cells is between 5 and 20% (see fig. 15). Some of these attach by cytoplasmic bridges to the wall of the blood vessel (Clark and Clark, '62) but others are free in the lumen. During regeneration, and especially during the first week, the number of paraldehyde-fuchsin positive cells increases enormously among both free and attached cells (figs. 15, 16A,B). Later the intensity of the paraldehyde-fuchsin staining decreases and the number of stainable granules decreases. Sometimes the cells appear to have secreted all their fuchsin positive material and many then have nuclei with very irregular surfaces, suggestive of a hyperactivity of the nucleus and the approaching exhaustion of the cells (fig. 16C). Although some paraldehyde-fuchsin positive blood cells appear in other vessels both in the stump and in the regenerate, this phenomenon is most pronounced in the dorsal blood vessel of the stump.

c. *Neurosecretory activity of the nerve cord*

The nerve cord in the most anterior segments of the amputated "tails," removed and fixed three days after the animals were collected, was compared with the nerve cord in the stumps of regenerating animals, fixed one to three weeks later, or 10 to 24 days after the animals were collected. A difference in the secretory activity of a single pair of neurons near the midline in the posterior third of each segment was observed. In the original tails the neurons stain only faintly if at all with paraldehyde-fuchsin, whereas in the stumps of one-week regenerates, they are intensely stained (fig. 17). The secretion granules are fine; the nuclei are large and often have a prominent nucleolus. It is

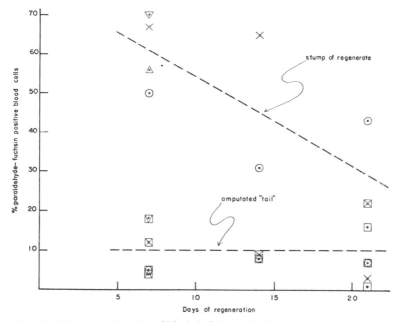

Fig. 15 Relative number of paraldehyde-fuchsin positive blood cells in the dorsal blood vessel of regenerating worms (▽ △ ✕ ○) compared with those in the amputated tails of the same worms (◫ ◩ ⊠ ⧆).

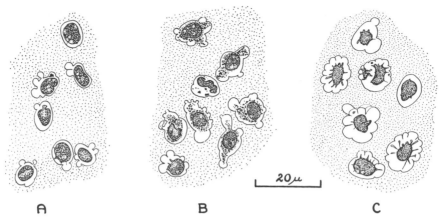

Fig. 16 Camera lucida drawings of blood cells in the dorsal blood vessel of A, an intact worm; B, a one-week regenerate; C, a three-week regenerate. Note numerous cytoplasmic granules in B, and "exhausted" nuclei in C.

sometimes possible to see secretion in the axon. The degree of staining is less in two- and three-week regenerates at which time the cytoplasm sometimes has several small or medium-size peripheral vacuoles, Cells in all segments of a given. animal show the same degree of neurosecretory activity regardless of their proximity to the wound.

Examination of intact animals which had been starved 17 and 24 days, however, revealed the same cells in a condition similar to that in regenerating worms. It is therefore possible that the secretory activity of these cells is a response to starvation rather than to regeneration, but this requires further data for confirmation.

d. *Involution of the gonads*

Comparison of the anterior segments of amputated "tails" with the posterior segments of the stumps of the corresponding animals after a period of regeneration revealed striking changes in the gonads. On the other hand the gonads of intact control worms starved for the same length of time appear to have undergone similar changes. Due to the impossibility of before and after comparisons in the latter case and to the wide variation in sexual development between individual animals, changes in the gonads are difficult to interpret, and

further study is required to distinguish between the effects of starvation and regeneration. For this reason, the changes observed will only be described briefly here.

The gonad of *Nephtys* arises from the coelomic epithelium overlying a complex of blood vessels in the anterior region of the neuropodium (Clark and Clark, '62). Sexual development begins with a medial-dorsal-wards outgrowth of new blood vessels, followed by multiplication of the cells of the coelomic epithelium ventrally. As primary gametocytes are formed they begin to accumulate around these newly formed vessels, but remain inclosed by coelomic epithelium until maturation is virtually complete. The germinal epithelium at this time consists primarily of small cells with nuclei having little or no distinctive chromatin granules and a small proportion of very large cells with characteristic heteropycnotic chromatin in the nuclei.

After a period of starvation and regeneration, the gonads show the following changes. There is a marked decrease in overall size, with a loss of gametocytes from the more dorsal region of the gonad where the blood vessels become bare. Cells in the intermediate stages of gametogenesis virtually disappear, the more mature ones apparently rapidly completing their devel-

96

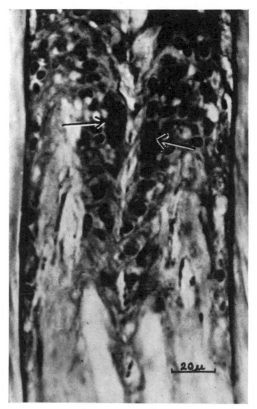

Fig. 17   Photomicrograph of paraldehyde-fuchsin positive neurosecretory cells (arrows) in ganglionic region of ventral nerve cord of 7-day regenerate.

opment and the less mature being resorbed. In males, a few mature sperm are produced and released into the coelom. In females, a few of the older oocytes in each segment develop nuclei of a type usually found only much later in development. The germinal epithelium undergoes a complete change, with the proportion of enlarged cells with heteropycnotic or partially heteropycnotic nuclei greatly increased. These cells show some mitotic activity.

Some of these changes are shown in figure 18. At the moment of amputation the gonads of this young male had devel-

oped to about one quarter of their final size and contained numerous primary and secondary spermatocytes (fig. 18A). Two weeks later the gonad at the same position in a stump segment shows complete regression, with only bare gonadial blood vessels remaining and a few mature sperm trapped in the coelomostome (fig. 18B).

DISCUSSION

In general, the sequence of events in regeneration in *Nephtys* is similar to that described in *Nereis diversicolor* (Herlant-Meewis and Nokin, '62). After healing, segments are formed in a manner identical to that in normal growth, except that the

97

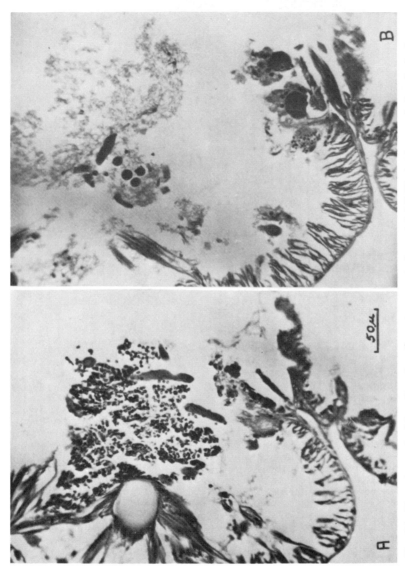

Fig. 18A Photomicrograph of frontal section of anterior segment of amputated "tail", showing gonad of moderate size sectioned at the level of the coelomostome (lower center) and composed mainly of primary and secondary spermatocytes.
Fig. 18B Photomicrograph of same region of posterior segment in stump of same worm after 14 days regeneration. Note the complete involution of the gonad and the occurrence of a few mature sperm in the coelomostome.

rate of addition of new segments is initially much more rapid and exceeds the rate of differentiation so that morphological differences between segments in regeneration are much less marked than between young segments in intact worms. This rapid initial rate of segment formation in *Nephtys* also agrees with that found in *Nereis* (Golding, '67b), although in the case of *Nereis* the peak rate occurs a few days later than in *Nephtys*. This discrepancy may be due to the fact that Golding used external criteria of segment formation, whereas in the present study internal indications of segmentation, which precede the formation of parapodia and other external features, were used.

The usefulness of purely descriptive studies such as the present one lies primarily in the new ideas they suggest to the investigator for further experimentation rather than in providing any absolute conclusions about such a dynamic process as regeneration. For example, such studies tell us little about the possible totipotence of blastema cells, about whether a visibly undifferentiated cell in fact has no specific function, nor about induction processes which may be occurring between tissues. The question of the source and fate of blastema cells in polychaetes, and in *Nephtys* in particular, has already been discussed extensively (Clark and Clark, '62), and only one new fact need be recorded here. In *Nephtys* there is some question whether the whole of the early accumulation of cells at the wound should be called a "regeneration blastema," since only a part of it actually participates in the regeneration of new tissue. These cells, located posteriorly near the new epidermis, reach the peak of their mitotic activity about the time that the non-dividing anterior zone which serves as an aid to wound healing is beginning to disperse.

Another observation from the present study is that wound healing must be virtually complete before new segments start to be formed. This is especially evident in worms where the wound area is enlarged due to extrusion of the gut and segmentation is delayed until an epidermal sleeve has grown over the exposed tissue. The last stage of healing occurs when the epidermal and endodermal basement membranes fuse, and it is at this junction, ventrally, that the new pygidium, always the first structure to regenerate and to visibly differentiate, is formed. Whether there is some causal relationship between the regeneration of the pygidium and the inception of segmentation remains to be investigated.

The mitotic counts used in this study to estimate the time of peak increment in cell numbers can only be taken as rough approximations since they are based not on the true mitotic index, but only on the total number of cells dividing in various areas of the regenerate. Despite the fact that worms of uniform size were used, variations in the total cell numbers in the regenerates undoubtedly add considerably to the error of the counts. Despite this limitation, however, when the data are expressed in terms of the number of segments regenerated it is evident that mitotic activity reaches its peak prior to the onset of segment formation. It is also apparent, simply by visual inspection of the relative size of the regenerate segments and subjective estimation of the nucleo-cytoplasmic ratios of the cells in the regenerate tissue that the growth of newly formed segments is dependent primarily on an increase in cell size, which occurs after the rates of mitoses and segment proliferation are significantly reduced. The small volume of a new segment means that the interrelations between various tissues, especially those connected by muscle fibers which later span great distances, can be established by single cells which are easily able to bridge the gaps at this time.

The nerve cord more than any other tissue exemplifies the fact that cell proliferation precedes growth of individual cells and enlargement of the segments. In fact, the relative size of the nerve cord, like that of the segmental blood vessels, is disproportionately large in the regenerate. This is especially true of the neuronal region, where the cell bodies are closely packed together in the young segments. It is possible that the peak of mitotic activity in the nerve cord is reached earlier than that of any other tissue and this raises the question of the role of the nervous system in regeneration. The general importance of the ventral nerve cord in polychaete regeneration has already been discussed (Clark and Clark, '62). It may be added here that in many animals, the nervous system is known to have a trophic effect on regeneration, although the exact mechanism by which this effect is exerted remains obscure (Singer, '65). In addition, it has been observed in *Nephtys* that along with the segmental blood vessels the first structures formed in a new segment are the gut suspensory muscles, which are soon followed by other segmental muscles. As noted earlier, except for the gut musculature, the muscles of *Nephtys* have at least one insertion on the epidermal basement membranes, and all but the gut suspensory muscles have both. It has recently been shown (Clark, '66) that at least some of the neuromotor end-plates appear to be located at these sites of muscle insertion on the basement membranes. Other segmental mesodermal derivatives, such as the gonads and nephridial complex, develop much later from coelomic epithelium associated with specific regions of the blood-vascular system. Epidermal derivatives, as chaetal sacs and sensory cells, also appear after a rudimentary segment is recognizable internally. It therefore appears likely that a new segment is initiated either by the terminal gut blood vessels and their associated fibroblasts or by cells in the ventral nerve cord, or possibly by an interaction between these two primordia. Kiortsis and Moraitou ('65) found that both the gut and the ventral nerve cord must be present for successful posterior regeneration in *Spirographis spallanzani*. On the other hand, the interesting observation of Holmes ('31) that in the absence of the nerve cord mesodermal segmentation was occasionally possible in posteriorly regenerating *Nereis virens* suggests that the gut and its associated structures may initiate a segment, whereas epidermal differentiation depends on the nervous system.

There is no indication of any neurosecretory activity in the regenerating nerve cord, as revealed by the paraldehyde-fuchsin technique, although there is a pair of cells with increased reactivity to this stain in each segment of the stump during the early weeks of regeneration. However,

as already pointed out it is not clear whether this increased reactivity is a result of regeneration or of starvation, since it appears to occur also in starved control animals.

The role of the gut in annelid regeneration appears to be variable. In oligochaetes, both Crowell ('37) and Kawakami ('61) have obtained successful posterior regeneration in the absence of endodermal tissue. And in the hesionid polychaete, *Magalia perarmata*, Abeloos ('50) has reported that new segments are regenerated posteriorly even when the gut is not present. On the other hand, Okada ('38) has demonstrated that for regeneration of more than just a pygidium, gut tissue is required in the syllid, *Autolytus edwarsi*, and, as just noted, Kiortsis and Moraitou ('65) found it necessary for posterior regeneration in *Spirographis*. However, in none of these studies has the role of the blood-vascular system associated with the gut been assessed, nor the possibility of collateral circulation examined in those cases where regeneration occurs in the absence of the gut. In *Nephtys*, the union of the gut and epidermis appears to be essential for the formation of a new pygidium and for segment delineation. In the latter function the blood vessels associated with the gut are of major importance while the endodermal tissue itself is passive. Such endodermal passivity has also been noted in the ariciid, *Scoloplos armiger* (Thouveny, '63). This is in contrast to the situation described in *Nereis massiliensis* (Sicard-Bruslé, '57) and *N. diversicolor* (Herlant-Meewis and Nokin, '62), where some endodermal basophilia is seen.

The enlarged blood vessels, particularly those associated with the gut during segment formation, require an increased blood volume in the regenerate zone. This must be supplied either by constriction of blood vessels elsewhere in the body, which has not been observed, or by production of new blood. The striking changes observed in the blood cells, especially those in the dorsal vessel of the stump, suggest these cells function in producing at least some of the blood constituents. Some of these cells appear to be attached to the ventral wall of the dorsal blood vessel and are probably analogous to the "heart body" described in the same vessel in various polychaetes (Cuénot, '91; Picton, '99). The significance of the increase in paraldehyde-fuchsin staining in these cells remains obscure, although the later decrease in stainable granules coupled with the bizarre appearance of the nuclei points to hyperactivity followed by exhaustion. The apparent increase in free blood cells observed in the early stages of regeneration in both *Nereis* (Herlant-Meewis and Nokin, '62) and *Nephtys* (Clark and Clark, '62) along with the temporary increase in paraldehyde-fuchsin stainability seen in *Nephtys* blood cells near the wound (Clark and Clark, '62) suggests that some of the activated cells originally attached to the walls of the dorsal vessel may be released into the circulation.

Another means of increasing blood volume locally would be by active transport of semi-permeable constituents of the coelomic fluid through the walls of the vessels into the blood. This possibility is suggested by the observation that the blood in newly formed vessels in the regenerate appears more granular and less dense than does blood elsewhere, indicating a lower concentration of high-molecular weight components such as hemoglobin.

In view of the hormonal role of the supraoesophageal ganglion in regeneration of new segments by *Nephtys* (Clark, Clark and Ruston, '62; Clark, '65) and other polychaetes (Herlant-Meewis, '64; Golding, '67a,b) it is interesting to speculate on the possible site of action of the neurosecretory material. Direct axonal transport to the wound is ruled out, at least in *Nereis,* in which ganglia transplanted to the coelom of decerebrate, caudally amputated worms are equally effective in inducing regeneration (Golding, '67a). In this connection, it should be emphasized that the recent work of Golding, ('67a,b) on *Nereis* indicates that the continuing presence of the hormone is required throughout the regenerative process and is not merely required to initiate segment regeneration. Moreover, although there is a minimum amount of hormone required, the normal rate of regeneration cannot be accelerated by the presence of excess amounts of hormone.

One possible target organ is the "heart body" of the dorsal blood vessel which may be partly responsible for the increased vascularity in the young regenerate. Clark and Evans ('61) noted that if the supraoesophageal ganglion is removed within four days after posterior amputation in *Nereis diversicolor,* vascularization fails to occur and no regeneration takes place. Another possible action of the "brain hormone" is the stimulation of somatic mitosis, as suggested by the work of Clark and Bonney ('60) and Clark ('65). When the ganglion is extirpated in nereid polychaetes, not only is regeneration inhibited, but precocious sexual maturation is induced (Durchon, '62). There seems to be a reciprocal balance between these two physiological states. Support for this appears in the present observation that rapid involution of half-mature gonads occurs during regeneration. It is necessary, however, to add a note of caution here, since it is possible that this involution is a result not of the regeneration but of the concommitant starvation which is the usual regimen during experiments on polychaete regeneration. In some oligochaetes, where wounding causes cessation of feeding, regression of gonads follows (Herlant-Meewis, '64). Moreover, fasting alone is sufficient to interrupt sexual development. In the latter instance certain neurosecretory cells in the ventral nerve cord become active, which is reminiscent of the observations made in the present study on *Nephtys.* It is also known that sexual maturity, and more especially a normal, well-fed nutritional state are inhibitory for caudal regeneration in the oligochaete *Eiseniella tetraedra* (Gay, '63). It thus appears possible that the capacity for either somatic or sexual development is dependent on nutritional state and that all three factors may be interacting by way of a system of hormonal checks and balances located in the central nervous system. Experimental investigation of the relation between nutritional and hormonal factors in polychaetes where decerebrate animals are unable to feed may prove difficult. However, significant amounts of organic material can be taken up across the body wall of polychaetes (Stephens, '63, '64; Stephens and Schinske, '61), and it may be possible to maintain well nourished experimental animals by dissolving appropriate nutrients in the ambient sea water.

LITERATURE CITED

Abeloos, M. 1950 Régénération postérieure chez *Magalia perarmata* Marion et Bobr. (Annélide Polychète Hésionide). C. R. Acad. Sci., Paris, 230: 477–478.

Clark, M. E. 1964 Biochemical studies on the coelomic fluid of *Nephtys hombergi* (Polychaeta: Nephtyidae), with observations on changes during different physiological states. Biol. Bull., Woods Hole. 127: 63–84.

Clark, M. E. 1965 Cellular aspects of regeneration in the polychaete *Nephtys*. In: Regeneration in Animals and Related Problems. (Ed. V. Kiortsis and H. A. L. Trampusch). pp. 240–249.

——— 1966 Histochemical localization of monoamines in the nervous system of the polychaete *Nephtys*. Proc. roy. Soc. B., 165: 308–325.

Clark, M. E. 1968 Free amino-acid levels in the coelomic fluid and body wall of polychaetes. Biol. Bull., Woods Hole, 134: 35–47.

Clark, M. E., and R. B. Clark 1962 Growth and regeneration in *Nephtys*. Zool. Jb. Physiol., 70: 24–90.

Clark, R. B. 1962 On the structure and function of polychaete septa. Proc. Zool. Soc. Lond., 138: 543–578.

Clark, R. B. 1964 Dynamics in Metazoan Evolution. (Clarendon; Oxford).

Clark, R. B. and D. G. Bonney 1960 Influence of the supra-oesophageal ganglion on posterior regeneration in *Nereis diversicolor*. J. Embryol. exp. Morph., 8: 112–118.

Clark, R. B., and S. M. Evans 1961 The effect of delayed brain extirpation and replacement on caudal regeneration in *Nereis diversicolor*. J. Embryol. exp. Morph., 9: 97–105.

Clark, R. B., M. E. Clark, and R J. G. Ruston 1962 The endocrinology of regeneration in some errant polychaetes. Proc. III. intern. Conf. Neurosecretion (Ed. H. Heller and R. B. Clark), Mem. Soc. Endocrinol., 12: 275–286.

Crowell, P. S. 1937 Factors affecting regeneration in the earthworm. J. exp. Zool., 76: 1–33.

Cuénot, L. 1891 Études sur le sang et les glandes lymphatiques dans la série animale (2ᵉ partie: invertébrés). Arch. Zool. exp. gén., 9: 365–475.

Durchon, M. 1962 Neurosecretion and hormonal control of reproduction in Annelida. In: Third International Symposium on Comparative Endocrinology, Oiso, Japan. Gen comp. Endocrinol. Suppl., 1: 227–240.

Gay, R. 1963 Influence de la maturité sexuelle et du jeûne sur la régénération caudale d' *Eiseniella tetraedra* F. typica (Sav.). Bull. Soc. zool. Fr., 88: 625–631.

Golding, D. W. 1967a Neurosecretion and regeneration in *Nereis*. I. Regeneration and the role of the supraoesophageal ganglion. Gen. Comp. Endocrinol., 8: 348–355.

Golding, D. W. 1967b Neurosecretion and regeneration in *Nereis*. II. The prolonged secretory activity of the supraoesophageal ganglion. Gen. comp. Endocrinol., 8: 356–367.

Herlant-Meewis, H. 1964 Regeneration in annelids. Adv. Morphogen., 4: 155–215.

Herlant-Meewis, H., and A. Nokin 1962 Cicatrisation et premiers stades de régénération pygidiale chez *Nereis diversicolor*. Ann. Soc. roy. Zool. Belg., 93: 137–154.

Holmes, G. E. 1931 The influence of the nervous system on regeneration in *Nereis virens* Sars. J. Exp. Zool., 60: 485–509.

Kawakami, I. K. 1961 Experimental analysis of factors influencing regeneration in the earthworm. Japan J. Zool., 13: 141–164.

Kiortsis, V., and M. Moraitou 1965 Factors of regeneration in *Spirographis spallanzanii*. In Regeneration in Animals and Related Problems. (Ed. V. Kiortsis and H. A. L. Trampusch), pp. 250–261.

Okada, Y. K. 1938 An internal factor controlling posterior regeneration in syllid polychaetes. J. mar. biol. Assoc. U. K., 23: 75–78.

Picton, L. J. 1899 On the heart body and coelomic fluid of certain polychaetes. Quart. J. micr. Sci., 41: 263–302.

Sicard-Bruslé, S. 1957 Source de l'histogenèse dans la régénération caudale de l'Annélide *Nereis massiliensis* Moquin-Tandon. C. R. Acad. Sci., Paris, 245: 1668–1669.

Singer, M. 1965 A theory of the trophic nervous control of amphibian limb regeneration, including a re-evaluation of quantitative nerve requirement. In: Regeneration in Animals and Related Problems. (Ed. V. Kiortsis and H. A. L. Trampusch), pp. 20–32.

Stephens, G. C. 1963 Uptake of organic material by aquatic invertebrates— I. Accumulation of amino acids by the bamboo worm, *Clymenella torquata*. Comp. Biochem. Physiol., 10: 191–202.

Stephens, G. C. 1964 Uptake of organic material by aquatic invertebrates. III. Uptake of glycine by brackish-water annelids. Biol. Bull., Woods Hole, 126: 150–162.

Stephens, G. C., and R. A. Schinske 1961 Uptake of amino acids by marine invertebrates. Limnol. and Oceanog., 6: 175–181.

Thouveny, Y. 1963 Sur les premières phases de la régénération postérieure de l'Aricien *Scoloplos armiger* O. F. M. Bull. Soc. Zool. France, 88: 86–94.

DEPOSITION OF CUTICULAR SUBSTANCES *IN VITRO* BY LEG REGENERATES FROM THE COCKROACH, *LEUCOPHAEA MADERAE* (F.)

E. P. MARKS AND R. A. LEOPOLD

The appearance of cuticle-like membranes on the surface of insect tissues maintained *in vitro* has been reported by several investigators. Demal (1956) found a thin membrane covering explants of leg imaginal discs from 3-hour-old *Drosophila* prepupae held *in vitro*. He also observed thickening of the cuticle and the formation of tarsal claws. Sengal and Mandaron (1969) explanted leg imaginal discs of late-instar *Drosophila* larvae with the cephalic complex (including the ring gland); when the cultures were treated with ecdysone, many leg discs showed morphogenic changes that included the appearance of setae, surface sculptpuring, and claws. Marks and Reinecke (1964) reported the secretion of a gelatinous material on the epidermis of leg regenerates of the cockroach, *Leucophaea maderae* (Fabr.), and Larsen (1966) cultured heart fragments from cockroach embryos and found a cuticle-like membrane covering vesicles that formed on these explants. Miciarelli Sbrenna, and Colombo (1967), working with explants of body wall from the abdomen of nymphs of the locust, *Schistocerca gregaria* (Forsk), found that the developmental stage of the donor insect greatly influenced the deposition of cuticle. Agui, Yagi, and Fukaya (1969) treated explants of epidermis from diapausing larvae of *Chilo suppressalis* (Walker) by adding small amounts of ecdysterone to the culture medium, and within 48 hours, the explant shed its old cuticle and deposited a new one.

Ritter and Bray (1968) cultured clots of blood from the cockroach, *Gramphadorina portentosa* in a medium that was free of insect plasma. The subcultures developed strands, flakes, and platelets around and within the clot, and the investigators concluded that the deposits were made up of a protein-chitin complex and that the complex was deposited on the surface of the culture vessel. However, the electron micrographic studies of Locke (1964) indicated that at least *in vivo*, the precursors are secreted by the epidermal cells.

Deposition of the cuticle in the regenerating leg of nymphs of the cockroach, *L. maderae*, was studied *in vivo* by Leopold and Marks (unpublished observations), and they found that two sheaths formed during regeneration. One appeared early in the development of the regenerate and thickened to form a gelatinous envelope around the regenerate. This layer apparently consisted of a protein-carbohydrate complex. The second sheath contained chitin, appeared late in development, thickened, and became wrinkled and rugose. The initial sheath was sloughed before molting, and the underlying chitin-bearing layer became the functional cuticle.

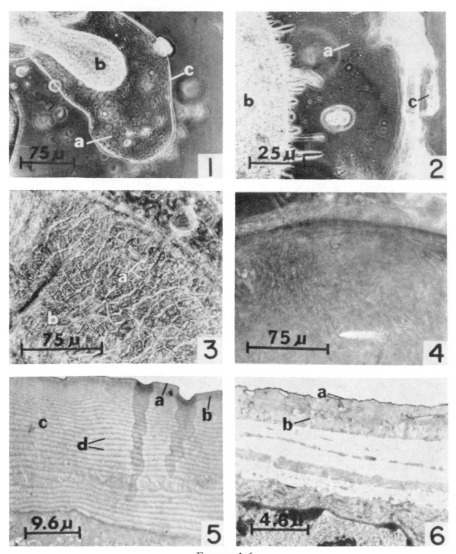

FIGURES 1–6.

FIGURE 1. A gelatinous sheath (a) laid down by the explant (b). The sharply defined outer border (c) is occasionally found in old, untreated cultures: *in vivo* 35 days: *in vitro* 53 days.

FIGURE 2. Gelatinous sheath (a) secreted by untreated explant (b) shows epithelial cells (c) extending into the sheath: *in vivo* 25 days; *in vitro* 40 days.

FIGURE 3. Detail of the surface of a mature leg regenerate that developed spontaneously *in vitro*. Seta (a) and surface sculpture (b) are characteristic: *in vivo* 35 days; *in vitro* 28 days.

In the present study, we have investigated some of the processes involved in cuticle secretion by combining the leg regenerate system with an *in vitro* methodology and histochemical analysis.

MATERIALS AND METHODS

Cockroach leg regenerates of different ages were prepared as described by Marks (1968). In the first tests, leg regenerates allowed to develop eight days after leg removal were dissected and placed under dialysis strips in Rose chambers with glass coverslips. The chemically defined M-7 medium was used. Six days later, they were examined with a phase contrast microscope, and the chambers were refilled with medium containing the test substance. After six more days, they were re-examined, and those showing cuticle-like deposits were scored as positive.

In the second test, leg regenerates were allowed to develop from 10 to 35 days after leg removal and were then dissected and examined. Those showing evidence of a cuticle were discarded, and the remainder were placed in Rose chambers between the dialysis strip and a glass coverslip. A plastic coverslip was added, and the chambers were filled with M-18 culture medium to which 5% fetal calf serum was added. The test was arranged with pairs of chambers, each containing a pair of leg regenerates from opposite sides of a single cockroach. One chamber of each pair was treated with ecdysterone, and the other was used as the control. Some of the controls were completely untreated, and others were treated with the inactive analog 22-iso-ecdysone (kindly supplied by Dr. John Siddall, Zoecon Corp., Palo Alto, California). The pairs were examined daily with a phase contrast microscope, and differences between the treated and untreated chambers were recorded. If the control showed evidence of developing cuticle, the treated chamber paired with it was removed from consideration. After 15 days, pairs in which the test leg developed a cuticular deposit were separated, and the test leg was discarded. Then the control was given 5 to 10 μg of ecdysterone to prove that its failure to produce cuticle had occurred because of the absence of the hormone and not from other causes.

The presence of chitin in selected specimens was verified by using the fluorescent enzyme technique of Benjaminson (1969). Frozen sections of the leg regenerate tissue were treated with chitinase conjugated with the fluorescent dye lissamine rhodamine B 200 chloride. They were examined by fluorescence microscopy, and fluorescence in cuticular structures was accepted as evidence of chitin.

RESULTS AND DISCUSSION

Deposition of the sheath

Leg regenerates that were allowed to develop for eight days before explanation normally showed no cuticle-like deposits when they were explanted. However,

---

FIGURE 4. Detail of the surface of a 10-day-old leg regenerate treated with 12 $\mu g/ml$ of ecdysterone. Setae are present but surface sculpture is absent: *in vivo* 10 days; *in vitro* 18 days.

FIGURE 5. Electron micrograph of cuticle from the leg of a newly molted cockroach. Epicuticle (a), sclerotized layer (b), and endocuticle (c) are clearly visible. Laminae (d) of the endocuticle are well formed and compact.

FIGURE 6. Electron micrograph of cuticle formed by a 25-day-old leg regenerate *in vitro* shows well-formed epicuticle (a) but poorly formed endocuticle (b): *in vivo* 25 days; *in vitro* 30 days.

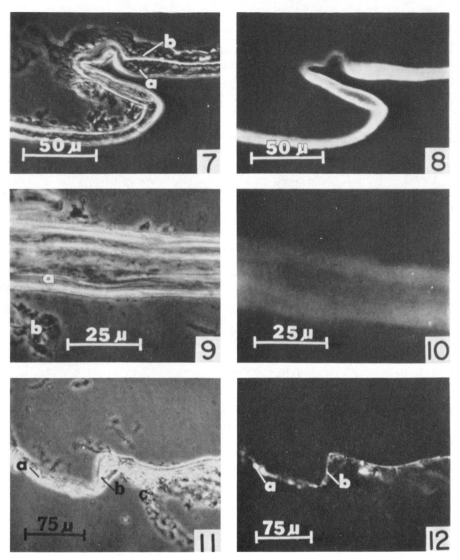

FIGURES 7–12.

FIGURE 7. A section through cuticle from leg of freshly molted cockroach. This specimen is shown under phase contrast illumination. Cuticle (a), which is refractible, is underlain by the epidermis (b).

FIGURE 8. Same specimen as figure 7 subjected to Benjaminson chitinase procedure and fluorescent microscopy. The chitinase-dye complex that has conjugated with the cuticle fluoresces. Note that the epidermis does not conjugate with the enzyme-dye complex and is no longer visible.

106

when they were held *in vitro,* 13 per cent produced deposits of gelatinous material within 14 days. The deposit appeared as a thin transparent layer that gradually thickened (Figs. 1 and 2); also, flakes and threads of darker material and occasional layering of this material appeared. As the cultures aged, the outer border of the layer often became sharply defined and yellowish (Fig. 1). No evidence of definitive cuticle appeared even though some cultures were held as much as 120 days. Fluorescent chitinase tests showed that the gelatinous layer, which probably represented the protective sheath seen in *in vivo* preparations, was devoid of chitin.

The protective role of this sheath, which may be related to wound cuticle (Sannasi, 1968), is supported by the finding that methanol (0.5 μl/ml of medium) and incubates of muscle tissue were as effective in inducing sheath deposition as were ecdysterone (2.5 μg/ml of medium) and incubates of the prothoracic gland. This sheath material is probably the same as the gelatinous substance reported by Marks and Reinecke (1964) and may also be the same as the cuticular material reported by Larsen (1966).

Deposition of the cuticle

The first evidence of the deposition of the cuticle *in vitro* was the appearance of a thin refractile layer between the epidermal cells and the protective sheath. As the deposition continued, the epidermal cells rounded up, contracted, and produced a pebbled appearance. Parallel ridges appeared and increased in size and number so the entire surface of the explant had a rugose appearance. Setae formed from hair-like trichogen cells that protruded from the explant, and the tormagen cell became embedded in the surface and sclerotized (Figs. 3 and 4). The deposition of the cuticle was usually terminated by the withdrawal of the epidermal cells from the secreted structures and the development of a space between them. This process apparently represents the *in vitro* counterpart of apolysis.

Examination of the ultrastructure showed that the cuticle regenerated *in vivo* possessed a thick endocuticle at the time of molting, and that the lamellae were compact and well formed (Fig. 5). Cuticle that formed *in vitro* (Fig. 6) showed a well developed epicuticle, but the endocuticle was only partially formed and not as compact as cuticle formed *in vivo;* separation of the cuticle from the epidermis was apparent.

FIGURE 9. Section through cuticle (a) formed by a 25-day-old leg regenerate *in vitro* seen with phase contrast illumination. Epidermis (b) has been largely stripped away in sectioning. The membrane is folded so the two layers are visible: *in vivo* 25 days; *in vitro* 20 days.

FIGURE 10. The same specimen as in Figure 9 subjected to the Benjaminson chitinase procedure and viewed by fluorescent microscopy shows two layers of fluorescence. Note fluorescence is not as intense as in Figure 8.

FIGURE 11. A section through cuticle formed *in vitro* by a 10-day-old leg regenerate viewed under phase contrast illumination. Note refractile droplets (a) appearing between the cuticle (b) and epidermis (c): *in vivo* 10 days; *in vitro* 21 days.

FIGURE 12. Same specimen as in Figure 11 subjected to the Benjaminson chitinase procedure and viewed by fluorescent microscopy. Both droplets (a) and solid cuticle (b) fluoresce, indicating the presence of chitin.

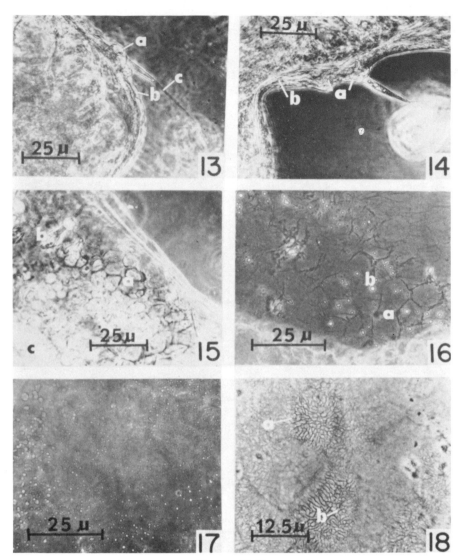

FIGURES 13–18.

FIGURE 13. A leg regenerate treated with 5 μg/ml of ecdysterone shows seta formation. Base of seta lies in the socket (a) that protrudes from the surface of the cuticle (b). The unsclerotized tip of the trichogen cell (c) is visible: *in vivo* 25 days; *in vitro* 25 days.

FIGURE 14. Another area of the same leg regenerate shown in Figure 13 shows two setae with well-formed sockets (a) embedded in cuticle (b). Only the base of the seta is sclerotized; upper end remains membranous: *in vivo* 25 days; *in vitro* 25 days.

FIGURE 15. Reticular pattern (a) and abnormal setae (b) are present on the surface of cuticle developing *in vitro*. Epidermis (c) has withdrawn, leaving the cuticle behind: *in vivo* 25 days; *in vitro* 25 days.

The dense cuticle formed *in vivo* demonstrated a strong reaction for chitin (Figs. 7 and 8) when it was tested by the fluorescent chitinase procedure. When the cuticle that formed *in vitro* was tested by this method, the reaction was weak but consistently positive (Figs. 9 and 10). The lower intensity of the fluorescence probably resulted from the lower density of the endocuticle in the *in vitro* preparations.

In some specimens, particularly among the younger explants, droplets of hyaline material often accumulated between the developing cuticle and the surface of the epithelial cells. When one such explant was sectioned and tested for chitin, we found a pebbled inner surface that contained chitin in the form of droplets (Figs. 11 and 12). Such droplets appeared only after treatment with ecdysterone and were most commonly found when the dosage was very low. The development of setae *in vitro* occurred with greatest frequency in explants that were taken from older nymphs (Table I). In some cases, the socket formed on the surface of the cuticle (Fig. 13) rather than imbedded in it (Fig. 14), and generally the sclerotization was incomplete.

We occasionally found cuticle forming over the surface of a vesicle, and when the vesicle later collapsed and withdrew the delicate epithelium, the remaining cuticle was left with a pattern representing the outline of the cells that deposited it (Figs. 15 and 16). The presence of the hyaline droplets often made it difficult to observe these patterns, but in those areas where they were absent, a pattern of ridges within the area laid down by a single cell was seen (Fig. 18). This apparently represented the imprint of the surface of the cell itself.

When leg regenerates were dissected late in the molting cycle (35 days), a large number had already initiated cuticle development *in vivo*. Those that had not were placed *in vitro*, and more than half continued their development without stimulation (Table I). Therefore, meaningful studies of the induction of cuticle development could only be made by using paired chambers containing legs from opposite sides of the same insect.

In an attempt to simplify our procedures and to increase our efficiency, we made a study of younger leg regenerates. Paired chambers were used, and leg regenerates from 5- to 30-day-old were treated with different doses of ecdysterone. The results indicated that while all doses above 2 μg/ml had about the same effect on cuticle production, the frequency with which the leg regenerates responded to stimulation increased with age (*in vivo*) until a 100 per cent response was reached at 20 days. There was also a corresponding increase in the frequency of seta formation with age.

The increase in ability to respond to the hormone with increased age seemed to continue after the tissue was placed *in vitro*. Thus, 10-day-old legs treated

FIGURE 16. Reticular pattern (a) is visible in this very thin cuticle. Droplets (b) probably containing chitin (Figure 11) are present. Epidermis has withdrawn: *in vivo* 25 days; *in vitro* 10 days.

FIGURE 17. Detail of the surface of a developing cuticle shows droplets accumulating between the cuticle and the epidermis. Droplets probably contain chitin; *in vivo* 10 days; *in vitro* 18 days.

FIGURE 18. Surface detail of developing cuticle shows polygonal outlines of cells (a). Within the outlines is a pattern (b) that represents the negative image of the irregular surface of the cell. This may represent the first step in the formation of the surface sculpture: *in vivo* 25 days; *in vitro* 10 days.

four days after explanation gave a 20 per cent (2/10) response, but when the water controls for this series were treated with ecdysone 14 days later, there was a 66 per cent (4/6) response. A similar but less marked effect was obtained with 15-day-old leg regenerates. The untreated controls did not produce cuticle.

A second test was set up to determine the minimum dose of ecdysterone that would produce a 100 per cent response in 25-day-old regenerates. Paired chambers were not used because regenerates of this age did not spontaneously produce cuticle under experimental conditions. In the 12 specimens treated with 2 μl of water and in the 10 specimens treated with 24 μg/ml of 22-iso-ecdysterone, there was no development of cuticle; all specimens developed cuticle when they were later treated with 10 μg/ml of ecdysterone for 25 hours. In the experimental chambers, doses of ecdysterone in water ranging from 2.5 μg/ml of nutrient to 0.05 μg/ml were given; then after seven days, the chambers were emptied and refilled with fresh nutrient, and the date when cuticle deposition

TABLE I

The effect of age at time of explanatation on development of cuticle in cockroach leg regenerates in vitro

Days after leg removal	Number tested	Per cent developing without treatment	Number (and per cent) developing after treatment with 2–10 μg/ml ecdysterone	
			Cuticle	Setae
10	10	0	2 (20)	1 (10)
15	8	0	7 (88)	1 (12)
20	6	0	6 (100)	2 (33)
25	16	0	16 (100)	10 (62)
30	4	0	4 (100)	3 (75)
35	5 (11)*	54	5 (100)	3 (60)

* Eleven were set up, and six (54%) developed spontaneously. All five remaining regenerates developed cuticle when treated with ecdysterone.

first appeared was recorded. A 100 per cent response was obtained with 2.5 μg/ml, a 90 per cent response was obtained with doses down to 0.5 μg/ml, a 37 per cent response was obtained with a dose of 0.2 μg/ml, and no response was obtained with a dose of 0.1 μg/ml or less. All the controls that were later exposed to the hormone developed cuticle.

The appearance of setae and surface sculpture and the presence of chitin indicated that the cuticular deposits formed *in vitro* in response to stimulation by ecdysterone are the same as those deposits that are present *in vivo* at the time of molting. Thus, it is likely that the structures reported by Demal (1956) and Sengel and Mandaron (1969) as appearing on imaginal leg discs taken from late-instar *Drosophila* larvae probably also contained chitin. In Demal's studies, development was spontaneous since stimulation occurred before the leg imaginal discs were explanted, but evaluation of the results of Sengel and Mandaron is more difficult because the brain and ring gland were present in the same cultures as the imaginal discs and because development occurred in some of the carrier-control cultures. Interaction between the glands and the ecdysone cannot be ruled out since Burdette, Hanley, and Grosch (1968) reported that the effect

110

of ecdysone on the ocular imaginal discs of *Drosophila* was enhanced by the presence of the cephalic lobes in the culture, and Williams (1952) showed that ecdysone can stimulate secretion by the prothoracic glands. However, the amount of ecdysone used by Sengel and Mandaron in the culture medium would probably have been sufficient to induce cuticle secretion by the leg imaginal discs, even in the absence of the gland explants. The work of Agui *et al.* (1969) further demonstrated that the entire process of molting, including the shedding of the old cuticle and the deposition of the new, can be induced in short-term cultures by adding ecdysterone to the culture medium.

In leg regenerates, the deposition of endocuticle *in vitro* is usually incomplete and varies considerably from one specimen to another. In general, the 35-day-old leg regenerates that produced cuticle *in vitro* tended to produce heavier deposits than did the 10- to 20-day-old ones. Two explanations for this are offered: The *in vitro* conditions may have been such that the normal synthesis of the endocuticle by the epithelial cells could not occur; therefore, the cuticle secreted *in vitro* would consist primarily of the epicuticle and that portion of the endocuticle secreted before the synthesis mechanism ceased to function. The other possibility, suggested by the experiments of Philogene and McFarlane (1967), is that the epithelial cells merely lay down a material that is either produced elsewhere or whose precursors are produced elsewhere. Then, by isolating the leg regenerate, we cut off the source of supply of this material except for that amount explanted along with the leg regenerate. This amount would be greater in regenerates explanted later in the molting cycle.

The fact that a 24-hour exposure of the leg regenerate to 10 μg/ml of ecdysterone was sufficient to induce cuticle deposition a week later suggests that the so-called "endogenous hormone" present in explants taken from insects late in the molting cycle probably refers to the physiological state of the tissue rather than to the actual presence of the hormone in the tissue. The fact that 10-day-old leg regenerates (which are little more than blood-filled sacs of epithelial tissue) can be induced to produce a seta-bearing cuticle with relatively small doses of ecdysterone indicates that this physiological state depends on the action of the hormone as well as on the age of the tissue. In *Leucophaea,* removal of a leg late in the stadium when the ecdysone titer is presumably high resulted in the formation of a cuticle-covered papilla, and regeneration of the leg resumed only after the subsequent molt. This *in vivo* response closely resembeled the response produced *in vitro* by treatment of 10-day-old leg regenerates with ecdysterone in which premature cuticle deposition eventually terminated morphogenesis.

However, if the leg was removed earlier in the cycle and allowed a longer period of development, a complete leg formed, and the approach of molting merely speeded up and completed the process. Similarly, Postlethwait and Schneiderman (1970) showed that when leg imaginal discs of *Drosophila* cultured *in vivo* were treated with large doses of ecdysterone, metamorphosis complete with formation of pupal cuticle occurred.

In our studies, we found that cuticle deposition can be stimulated by small doses of hormone presumably because other tissues (blood and fat body) are not present to inactivate the residual hormone and the target tissue is thus exposed to low levels of the hormone over a period of several days. In *in vivo* cultures, very large doses of the hormone were required to produce such an effect, thus suggesting that there may be a relationship between the amount of hormone present and the length of time that the tissue is exposed to it.

Ohtaki, Milkman and Williams (1968) proposed that tissue competence results from the accumulation of physiological events caused by continuous exposure to a low titer of hormone over a long period. It is just such a situation that we reproduced *in vitro, i.e.,* a low dose (0.2 μg/ml) of hormone left in contact with the target organ over a long period (7 days). Our results tend to substantiate the hypothesis of Williams' group. Furthermore, it should be possible to use this experimental system as a basis for a series of time-dosage studies that would shed additional light on this question.

The authors acknowledge their indebtedness to J. G. Riemann for the electron micrographs in Figures 5 and 6 and to J. D. Johnson for preparing the chitinase-dye conjugate used in the Benjaminson technique. Special acknowledgment goes to Herbert Oberlander and A. Glenn Richards for their helpful comments and suggestions.

This work was supported in part by a grant from the Kales Foundation.

LITERATURE CITED

AGUI, N., S. YAGI AND M. FUKAYA, 1969. Induction of moulting of cultivated integument taken from a diapausing rice stem borer larva in the presence of ecdysterone. *Appl. Entomol. Zool.,* **4**: 156–157.

BENJAMINSON, M. A., 1969. Conjugates of chitinase with fluorescine isothiocyanate or lissamine rhodamine as specific stains for chitin *in situ. Stain Tech.,* **44**: 27–31.

BURDETTE, W. J. E. W. HANLEY AND J. GROSCH, 1968. The effect of ecdysones on the maintenance and development of ocular imaginal discs *in vitro. Tex. Rep. Biol. Med.* **26**: 173–180.

DEMAL, J., 1965. Culture *in vitro* d'ebauches imaginales de Dipteres. *Ann. Sci. Natur. Zool. Biol. Anim.,* **18**: 155–161.

LARSEN, W., 1966. Growth in an insect organ culture. *J. Insect Physiol.,* **13**: 613–619.

LOCKE, M., 1964. The structure and function of the integument in insects. Pages 370–470 in M. Rockstein, Ed., *The Physiology of Insecta, Volume III.* Academic Press, New York.

MARKS, E. P., 1968. Regenerating tissues from the cockroach, *Leucophaea maderae:* The effect of humoral stimulation *in vitro. Gen. Comp. Endocrinol.,* **11**: 31–42.

MARKS, E. P., AND J. P. REINECKE, 1964. Regenerating tissues from the cockroach leg: A system for studying *in vitro*. *Science,* **143**: 961–963.

MICIARELLI, A., G. SBRENNA AND G. COLOMBO, 1967. Experiments of *in vitro* cultures of larval epiderm of desert locust. *Experientia,* **23**: 64–65.

OHTAKI, T., R. MILKMAN AND C. WILLIAMS, 1968. Dynamics of ecdysone secretion and action in the fleshfly, *Saracophaea peregrina. Biol. Bull.,* **135**: 322–334.

PHILOGENE, B., AND J. McFARLANE, 1967. The formation of the cuticle in the house cricket. *Acheta domestica,* and the role of oenocytes. *Can. J. Zool.,* **45**: 181–190.

POSTLETHWAIT, J. H., AND H. SCHNEIDERMAN, 1970. Induction of metamorphosis by ecdysone analogues: *Drosophila* imaginal discs cultered *in vivo. Biol. Bull.,* **138**: 47–55.

RITTER, H., AND M. BRAY, 1968. Chitin synthesis in cultivated cockroach blood. *J. Insect Physiol.,* **14**: 361–366.

SANNASI, A., 1968. Hyaluronic acid in the scar tissue of cockroach. *Zool. Jahrb. Abt. All. Zool. Physiol. Tiere,* **74**: 319–327.

SENGAL, P., AND P. MANDRARON, 1969. Aspectes morphologiques du developpement *in vitro* des disques imaginaux de la *Drosophile. C. R. Acad. Sci. Paris,* **268**: 405–407.

WILLIAMS, C. M., 1952. Physiology of insect diapause IV. The brain and prothoracic glands as an endocrine system in the Cecropia silkworm. *Biol. Bull.* **103**: 120–138.

Regenerating Tissues of the Cockroach *Leucophaea maderae:* Effects of Humoral Stimulation *in Vitro*

EDWIN P. MARKS

When the leg of a cockroach nymph is removed at the trochantero-femoral joint during the first half of the molting period, regeneration of the missing limb occurs. This process starts in the base of the trochanter, and the developing leg extends upward into the coxa, growing at the expense of the mass of muscle already present. By the end of the intermolt period, a completely formed leg is present that is exposed at the subsequent molt. This process has been described in detail by Bodenstein (1955) for the cockroach, *Periplaneta americana* L. If the leg is removed during the latter half of the intermolt period, wound healing occurs but regeneration is delayed until the next intermolt period. This phenomenon was reported by O'Farrell and Stock (1953) for the cockroach, *Blattella germanica* L. Development of the leg regenerate is thus clearly under the control of the molting cycle of the insect; the molting cycle, in turn, is triggered by the secretion of the prothoracic gland.

In an earlier study (Marks and Reinecke, 1964), it was demonstrated that if leg regenerates from fifth instar nymphs of the cockroach *Leucophaea maderae* F. were removed at an early stage of development

and placed in a chemically defined nutrient medium, they survived for a considerable period and underwent a series of morphogenic changes. While it was difficult to correlate the activities observed *in vitro* with those that occur *in vivo,* many of the morphogenic events appeared to be similar.

Wigglesworth (1959) demonstrated that the main activity after wounding is cell migration; by this process, the wounded area is quickly covered by an epithelial sheet. In the stump of a freshly amputated leg, the closure of the wound thus provides an epithelial sac within which the leg regenerate, growing at the expense of the muscle tissue of the leg stump, will be formed. A second wave of cell activity then results in an aggregation of cells around the cut end of the fifth mesothoracic nerve, producing a small mass of essentially non-differentiated tissue that is present on the fifth day after leg removal. This tissue mass enlarges and differentiates rapidly until, by the eighth day, it consists of a sac of epithelial tissue enclosed (along with the remaining muscle tissue of the leg stump) within the epithelial lining of the coxa and trochanter (Fig. 1).

When the leg regenerate is explanted at

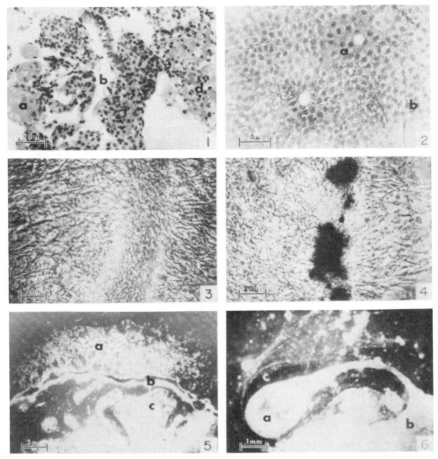

Fɪɢ. 1. Section through a portion of an 8-day leg regenerate from a nymph of the cockroach *Leucophaea maderae* F. showing: (a) 5th mesothoracic nerve; (b) cavity within the regenerate; (c) developing epithelium of regenerate; (d) old muscle tissue under attack by phagocytes.

Fɪɢ. 2. Surface of a vesicle growing on a leg regenerate. Two cell types are apparent:(a) fiber-like cells form network; (b) epithelioid cells form sheet.

Fɪɢ. 3. Migration of cells from a leg regenerate forms a thickened band of cells in the cell sheet that partially surrounds the explant.

Fɪɢ. 4. Red-brown pigment is deposited in the thickened band, and the cells on the side farthest from the explant (a) begin to degenerate.

Fɪɢ. 5. Fully mature leg regenerate showing: (a) dead cells outside barrier; (b) barrier impregnated with pigments secreted by the cells; (c) differentiating tissue inside protective barrier.

Fɪɢ. 6. Mature vesicle (a) grown from surface of explant (b) and covered with secreted cuticular material (c).

115

this time into a nutrient medium, there are usually two waves of cell migration to the supporting surface. The first wave of cells occurs within 2 days after explantation, and they live only 2 to 3 days and then die. They appear to be plasmatocytes from the hemolymph. By the end of 7 days, a second wave of migration occurs, frequently forming a more or less well-defined layer of epithelium that often remains in good condition for as long as 30 days after explantation. This sequence is similar to that reported by Day (1951) for wound healing in *Periplaneta* in which the first wave of migrating cells forms the wound plug while the second wave replaces the damaged epithelium.

In many cases, the sheet of cells produced undergoes a remarkable transformation in which a dense band of cells is formed that partially or completely surrounds the explant (Fig. 3). A red-brown pigment is slowly deposited in this band, starting as patches but eventually forming a complete barrier (Fig. 4). The portion of the cell sheet outside this barrier then degenerates while the portion inside the barrier participates in the formation of tissue buds and vesicles (Fig. 5). This process of cell migration and barrier formation closely resembles the wound healing process found in the closing of the cut leg stump of *Leucophaea*. The processes occurring (vesicle and tissue bud formation) in the tissue within the barrier are more similar to those involved in the formation of the leg regenerate itself (Fig. 6).

The formation of vesicles (hollow spheres filled with fluid) in cultures of embryonic tissues of various insects has been reported by a number of workers. Such vesicles may form on or within the existing tissue mass and bulge outward or they may arise *de novo* within a sheet of cells apart from the central tissue mass. The epithelium of the vesicle is made up of a web of fibroblast-like cells invested with a sheet of epithelioid cells (Fig. 2). The inflation of the vesicle with fluid is an active process; vesicles that develop lesions and thus deflate often seal up the hole and reinflate. The death of the cells, however, results in rapid and irreversible collapse. In *Leucophaea,* vesicles frequently produce a gelatinous secretion at the surface that closely resembles the early stages of cuticle formation in the intact cockroach. Larsen (1967) reported a similar cuticular sheath on the surface of vesicles forming on fragments of *Blaberus* embryos maintained *in vitro*. In some of the older leg-regenerate cultures, virtually all the living tissue eventually becomes involved in the formation of large and elaborate vesicles. While vesicles, as such, do not occur *in vivo*, there are periods early in the regeneration process when the leg regenerate consists essentially of a sac of epithelial cells (Fig. 1). In one case, when an entire early regenerate was explanted, the epithelial sac, open at the base, moved a constant stream of fluid for a period of several days through the wall and out of the opening (Marks and Reinecke, 1964). It would appear, then, that while vesicles formed *in vitro* and the leg regenerates formed *in vivo* are certainly not identical structures, the processes that are involved in their formation (epithelium formation and the movement of fluid to inflate the structure) are probably the same.

In a subsequent study, Marks and Reinecke (1965a) reported that the occurrence of these morphogenic activities was influenced by the presence of certain endocrine glands explanted into the culture chamber along with the regenerate tissue. It was also shown that the effect produced was dependent upon both the age of the regenerate tissue and the age of the glands. This effect was maximal when the regenerate was explanted 7 to 9 days after removal of the leg. Under these conditions, changes in the growth pattern of the leg regenerates became identifiable after 5 to 6 days *in vitro*.

Three problems were encountered in evaluating the results of these early experiments. One of them was that a small percentage of the controls either died or simply failed to develop, probably because of injury during explantation. This spontaneous degeneration was difficult to distinguish from atrophy caused by the pres-

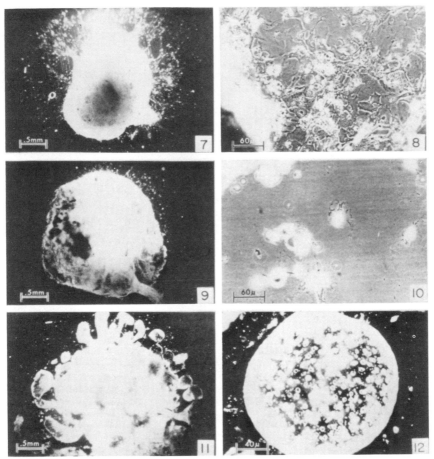

FIG. 7. Leg regenerate showing extensive undifferentiated outgrowth of cells.
FIG. 8. Detail of undifferentiated cell sheet.
FIG. 9. Leg regenerate showing typical atrophy of migrating cells.
FIG. 10. Detail of atrophied cells.
FIG. 11. Leg regenerate showing vesicle formation.
FIG. 12. Detail of detached vesicle.

ence of the glands. A second question was whether the products of the glands were affecting the leg regenerate directly or whether interaction was occurring, i.e., the products of one gland controlled the activity of another gland, which in turn produced substances that acted upon the leg regenerate. A third problem was that cell bridges often formed between the various explants. In such cases, interaction could have been induced by physical rather than humoral stimulation. The present study was undertaken to answer these questions and to demonstrate, if possible, the nature of the interactions among the various endocrine gland explants.

MATERIALS AND METHODS

Fifth instar nymphs of the cockroach, *L. maderae*, were isolated from the colony immediately after molting. The next day, the two middle legs were removed at the trochantero-femoral joint. Eight days later, the entire coxa was removed, and the mass of regenerating tissue was dissected. The isolated leg regenerates were placed in Rose multipurpose tissue chambers under dialysis strips, three to four regenerates in each chamber. One milliliter of nutrient medium (Marks and Reinecke, 1965b) was added, and the chambers were incubated at 26°C. Since Scharrer (1948) reported a burst of mitotic activity in the prothoracic glands of this insect as early as 6 days postmolt, they are presumably active at this time. The endocrine glands to be tested were therefore removed simultaneously with the leg regenerates (9 days into the fifth intermolt period, which, in this laboratory, usually lasts about 45 days). The glands were arranged into various combinations for testing, placed in tubes containing 1 ml of nutrient medium, and then incubated at 26°C for 6 days.

Two types of gland incubates were prepared. First, combinations of different glands were incubated in the same tube, permitting the glands to interact. Secondly, different kinds of glands were incubated in separate tubes and the incubates combined prior to injection into the chamber. This effectively prevented any interaction from occurring among the glands themselves. After incubating for 6 days, the leg regenerate tissues were examined and the spontaneous occurrence of new growth (Figs. 7, 8), cell atrophy (Figs. 9, 10), and vesicles (Figs. 11, 12) were recorded. The chambers were then treated with the gland incubates, incubated for 6 additional days, and then examined again. Any visible changes that occurred between the two readings were recorded. With this method, cell atrophy could be distinguished from spontaneous degeneration caused by damage at the time of explantation since the latter showed up plainly at the first reading; such damaged explants were eliminated. Vesicles formed on some leg regenerates spontaneously during the first 6-day period; other leg regenerates showed only cell migration. If vesicles were absent in the initial reading but appeared before the final reading, it was recorded as vesicle gain; if vesicles were present on an explant at the initial reading and disappeared by the final reading, it was recorded as vesicle loss. A minimum of 10 leg regenerates showing each type of development was tested for each gland combination tested.

RESULTS AND DISCUSSION

Effect Produced by Incubates Derived from Varying Numbers of Prothoracic Gland Pairs

When the incubate derived from various numbers of 8-day prothoracic gland pairs was tested on leg regenerates of the same age (Table 1), it was found that an increase in the number of glands incubated from one-half to two pairs resulted in an increase in the occurrence of both vesicle loss and cell atrophy. It was also observed

TABLE 1
REACTION OF LEG REGENERATES TO INCUBATES
OF DIFFERENT NUMBERS OF PROTHORACIC
GLAND PAIRS

No. of gland pairs incubated	Ratio of regenerates showing effect to number treated		
	Vesicle loss	Vesicle gain	Atrophy
None	4/23	5/28	2/47
1/2	2/10	2/12	1/22
1	4/11	5/16	1/27
1–1/2	7/18	3/13	7/31
2	8/19	0/18	16/32
3	6/14	1/10	9/21

that vesicle gain rose to a maximum between one and one and a half gland pairs and then dropped again. When the number of gland pairs incubated was increased from two to three, no further changes in response were found. When these results were converted to percentage of occurrence and expressed graphically, each of the three variables showed a different threshold of response to the prothoracic gland incubate (Fig. 13).

There appears to be little doubt that the responses of the leg regenerates to the presence of prothoracic gland incubate are caused by unidentified substances diffusing out of these glands into the culture medium. Interpreting these responses in terms of the events occurring *in vivo* is difficult, but a number of apparent similarities do occur. From a level *in vitro* of zero to one-half gland pair equivalents, there is little change

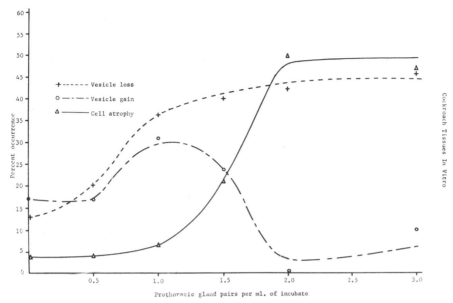

Cockroach Tissues In Vitro

FIG. 13. Effect of prothoracic gland incubate on cockroach leg regenerates *in vitro*.

except for a slight rise in vesicle loss. Since vesicle loss is frequently associated with vigorous cell migration at this level (and cell atrophy is virtually absent), it is suggested that this represents the first response to wounding, which is closure of the wound. Between one-half and one gland pair equivalents, both vesicle gain and vesicle loss showed a distinct increase. Since both may occur in the same chamber at the same time, this difference in response to the identical dose of gland incubate evidently is due to the physiological condition of the explant itself. Since both cell migration and vesicle formation appear to be involved in the regeneration process *in vivo*, it is not difficult to see how slight variations in age and physiological condition of the explant could alter the balance between these processes.

A sharp rise in cell atrophy accompanied by a similar drop in vesicle formation and a leveling off in vesicle loss occurred as the dosage was increased from one to two gland pair equivalents. The decrease in vesicle formation that occurred under these cir-

cumstances appeared to be a decrease in the ability of the cells to maintain the integrity of the membrane and a decrease in the transport of fluid through the membrane; these were apparently related to the increase in cell atrophy. Since most of the cells that migrated out of the explant died under these conditions, vesicle loss at this dosage remained at a high level.

In an earlier study, Marks and Reinecke (1965a) reported obtaining a response similar to that obtained with two gland pairs with a single pair of prothoracic glands explanted from cockroaches 18 to 21 days postmolt, or roughly half way through the intermolt period. O'Farrell and Stock (1953) reported that in *Blattella*, leg regeneration usually failed to occur until after the subsequent molt when the legs were removed after the middle of the first stadium. It thus appears that the increasing level of prothoracic gland output that stops vesicle formation and increases cell atrophy *in vitro* also inhibits the development of a leg regenerate *in vivo*.

While the evidence presented here is

largely circumstantial, the *in vivo* findings are in close enough agreement with the *in vitro* findings to suggest that the diffusable substance in question is related to the molting cycle and thus to the endocrine mechanism that controls it.

Effect of Incubates of Brain and Corpus Cardiacum on Prothoracic Gland

The brain and corpus cardiacum are known to produce stimulation of the prothoracic glands *in vivo*. Having found a method for approximating the output of the prothoracic gland *in vitro*, it became possible to investigate the effect on the prothoracic gland of incubates derived from the brain and corpus cardiacum. The results obtained from such experiments are reported in Table 2.

When a single pair of prothoracic glands was incubated in the presence of a single corpus cardiacum, the effect of the resulting incubate on leg regenerates gave a result approximately equal to that produced by one and a half prothoracic gland pairs alone. When a pair of prothoracic glands was incubated with a brain, however, the effect approximated that produced by half a pair of prothoracic glands incubated alone. When the number of prothoracic gland pairs incubated with the brain was increased, it was found that with the brain present, these glands still produced only about 50% of the effect produced when a similar number of prothoracic gland pairs were incubated alone. When a brain and prothoracic gland pair were incubated sepa-

rately and the incubates combined, the effect was closer to that produced by a single prothoracic gland pair alone.

The increase in activity produced by the prothoracic gland-corpus cardiacum incubate was somewhat less than expected in the light of the earlier finding of a high level of activity with 18- to 21-day-old glands (Marks and Reinecke, 1965b), in which presumably activation by the brain and corpus cardiacum had taken place *in vivo* before explantation. Two possible explanations for this come to mind: (1) At this relatively early stage of the intermolt period, only a low concentration of prothoracotropic hormone has accumulated in the corpus cardiacum; (2) explantation

TABLE 2

THE EFFECT PRODUCED WHEN BRAIN AND CORPUS CARDIACUM ARE INCUBATED
WITH THE PROTHORACIC GLAND

| | Ratio of regenerates showing effect to number treated | | | | | |
| | Vesicle loss | | Vesicle gain | | Atrophy | |
Glands	Ratio	%	Ratio	%	Ratio	%
Incubated together						
Corpus cardiacum + 1 pair prothoracic glands	3/11	27	3/11	27	3/19	15
Brain + 1 pair prothoracic glands	3/11	27	2/13	15	1/25	4
Brain + 2 pair prothoracic glands	3/13	23	4/11	35	2/25	8
Brain + 3 pair prothoracic glands	3/10	30	1/10	10	5/19	26
Incubated separately						
Brain + 1 pair prothoracic glands	4/12	33	3/12	25	0/24	0

may have damaged the release-receptor mechanisms so that the stimulating effect of this hormone was reduced.

More surprising was the apparent inhibition of the prothoracic gland when it was incubated with the brain. At first these results appeared to conflict with those of the earlier study when the results obtained with glands of several different ages had been lumped. When these results were separated according to the age of the gland, however, they agreed closely with those obtained in the present study in which a single brain produced 50% reduction of output, regardless of the number of prothoracic gland pairs used. Since this inhibition was not evident when the glands were incubated separately, it appears that some kind of gland interaction is occurring.

In an attempt to explain these rather

puzzling findings, three situations seemed possible. (1) During the period of incubation, the relatively large mass of brain tissue might be producing toxic metabolic products and/or using up nutrient substances and thus inhibiting the activity of the prothoracic gland. (2) The prothoracic gland may have been functioning normally but its products were taken up and either destroyed or retained by the brain tissue, making them unavailable to the leg regenerate. (3) The third suggestion arose from the fact that the inhibition percentage remained the same regardless of the number of prothoracic gland pairs employed, i.e., the brain was producing some type of regulatory substance that inhibited the activity of the prothoracic gland. In order to distinguish between these three possibilities, a series of experiments was carried out.

Effects of Incubates of Prothoracic Ganglion on Prothoracic Gland

When the prothoracic ganglion was incubated with a single prothoracic gland pair, the effect on the leg regenerates was nearly twice that produced by a single prothoracic gland pair alone. When these same glands were incubated separately, the result was approximately the same as that produced by a single gland pair alone. Thus, the presence of the prothoracic ganglion produced a stimulatory rather than a depressing effect on the activity of the prothoracic gland.

The foregoing experiment demonstrates that the effect of the brain in producing a depression in prothoracic gland output is not merely a matter of a large mass of nerve tissue interfering with the normal output of the prothoracic gland since the prothoracic ganglion, a nerve mass roughly equal in size to the brain, produced precisely the opposite effect. That tissue-breakdown products and medium exhaustion are not determining factors is evident since there is no significant effect when the same tissues (brain and ganglion) are incubated separately and the incubates combined. Since the prothoracic ganglion, like the other ganglia of the ventral nerve chain, contains what appears to be neurosecretory cells, it is assumed that the other ganglia would produce a similar effect. The finding that a diffusable substance produced by a ganglion of the ventral nerve chain is capable of increasing the output of the prothoracic gland recalls the work of Mc-Daniel and Berry (1957) who confirmed the findings of earlier workers that the prothoracic glands of *Antheraea* pupae could be activated by injury to the integument of the abdomen in the absence of the brain. Presumably, the activation of the prothoracic glands in this case is mediated by diffusible substances from the ventral nerve chain.

Of the three hypotheses proposed earlier to account for the depression in prothoracic gland activity produced by the presence of the brain, only one remains; namely, that a diffusible substance was being produced by the brain that depressed the secretory activity of the prothoracic glands. To pursue this hypothesis further, a fourth set of experiments was undertaken to determine whether the presence of the brain would inhibit the action on leg regenerates of a single prothoracic gland that had been stimulated to a higher level of output.

Effects of Incubates of Corpus Allatum-Cardiacum on Prothoracic Gland

To obtain corpora allata completely free from tissue of the corpus cardiacum proved to be difficult and not practical under the experimental conditions involved. Thus, the entire corpus allatum-cardiacum complex was tested. In the present study, when the allatum-cardiacum complex was incubated with a single prothoracic gland pair, the output of the prothoracic gland was increased some 50% over that obtained with the cardiacum and 100% over that obtained with the prothoracic gland alone (Table 3). When the same glands were incubated separately and the incubates combined, this effect was longer apparent, indicating that gland interaction was occurring. When a brain was incubated with the allatum-cardiacum complex and a prothoracic gland pair, the effect was roughly the same as that produced by one and a half prothoracic gland pairs alone, or a

TABLE 3
THE EFFECT PRODUCED BY THE CORPUS ALLATUM-CARDIACUM COMPLEX AND
PROTHORACIC GANGLION ON THE ACTIVITY OF THE PROTHORACIC GLAND

Gland	Ratio of regenerates showing effect to number treated					
	Vesicle loss		Vesicle gain		Atrophy	
	Ratio	%	Ratio	%	Ratio	%
Incubated separately						
Corpus allatum-cardiacum + prothoracic gland pair	3/12	25	2/15	13	5/24	21
Prothoracic ganglion + prothoracic gland pair	4/12	33	2/11	18	0/23	0
Incubated together						
Corpus allatum-cardiacum + prothoracic gland pair	5/13	38	0/13	0	15/22	68
Corpus allatum-cardiacum + brain + prothoracic gland pair	7/15	46	2/13	14	4/27	14
Prothoracic ganglion + prothoracic gland pair	6/18	33	1/16	6	8/24	33

50% inhibition of the activity obtained with the allatum-cardiacum complex and prothoracic gland alone.

From these experiments, it appears obvious that the intact allatum-cardiacum complex produces roughly twice the effect on the prothoracic glands as that produced by the corpus cardiacum alone. Since Ichikawa and Nishitsutsuji-Uwo (1959) reported the occurrence of stimulation of the prothoracic gland by the corpora allata in immature insects and since Gilbert and Schneiderman (1959) reported that juvenile hormone extracts stimulated the prothoracic gland in vivo, it appears probable that the increased stimulation is caused by secretions from the corpora allata. This 50% increase in prothoracic gland output does not appear when the brain is present, and the over-all effect in this latter case is only slightly greater than that produced by a brain incubated with two pairs of prothoracic glands in the previous experiment.

Virtually the same effect was found in the 1964 series of experiments. When explanted with the allatum-cardiacum complex, one pair of prothoracic glands (7–21 days postmolt) produced atrophy in 66% of the cases; when a brain was added to this combination, the occurrence of atrophy dropped to 18%. Thus, the effect of the brain in decreasing the activity of the prothoracic gland is essentially the same whether the output is increased by increasing the number of glands (7-day) that are secreting at a relatively low level; whether the single (7-day) gland pair is stimulated by the presence of the allatum-cardiacum complex in vitro; or whether the gland is activated in vivo and explanted when its level of activity is already high.

Effects of Nonendocrine Tissues on Prothoracic Gland

Having demonstrated the presence of interaction between known endocrine tissues and the prothoracic gland by means of the effect of the incubates on leg regenerates, the question arose as to whether other tissues produced similar effects on the prothoracic glands.

To answer this question, a series of experiments was run. In the first, pieces of muscle approximating 1 mm^3 in size were incubated with one and a half prothoracic gland pairs, and the incubate was tested on 8-day-old leg regenerates. The results are given in Table 4. Vesicle loss and gain were virtually identical with those produced by one and a half prothoracic gland pairs alone, and toxicity was slightly decreased. When compared with the effects produced when prothoracic glands were incubated with tissues of known neuroendocrine activity, there was no evidence of interaction. The results compared most closely with those produced when a prothoracic ganglion and a single prothoracic

TABLE 4
THE REACTION OF LEG REGENERATES TO INCUBATES OF VARIOUS TISSUES

| | Ratio of regenerates showing the effect to the number treated | | | | | |
| | Vesicle loss | | Vesicle gain | | Atrophy | |
Tissues incubated	Ratio	%	Ratio	%	Ratio	%
Muscle + 1½ prothoracic gland pairs	6/13	46	2/12	17	2/19	10
Regenerate + 1 prothoracic gland pair	2/14	14	3/10	30	2/20	10
2 prothoracic ganglia	5/10	50	2/19	10	3/29	10

gland pair (a roughly equivalent amount of tissue) were incubated separately. When compared with the results expected if one and a half pairs of prothoracic glands were incubated with a brain, the high vesicle loss and atrophy and a low vesicle gain gave no indication of any effect on prothoracic gland output.

The loss of vesicles, as indicated earlier, can be caused by two factors: Toxicity that kills the cells causes disruption of the vesicle. Loss of morphogenic activity also occurs when cells migrate out of the vesicles and form monolayers on the glass surface. This type of activity was evident in the case in which two prothoracic ganglia were incubated together and the incubate tested; a 50% vesicle loss accompanied by only 10% cell atrophy resulted. Since this latter incubate also produced the wound healing–like phenomenon described earlier (38% of the cases), it is assumed that a high vesicle loss coupled with a low atrophy represents a tendency of the test tissues to produce wound healing rather than regenerate-forming activities.

In a second experiment, an 8-day-old leg regenerate was incubated with a single prothoracic gland pair of the same age, and the resulting incubate was tested for activity. The results are given in Table 4. The vesicle gain and atrophy produced were very close to those produced by a single pair of prothoracic glands alone. The vesicle loss, however, was near the level obtained when no glands of any kind were added. Thus, the presence of the leg regenerate in the incubate appeared to have no effect except on this one parameter. The high level of vesicle formation coupled with the low levels of vesicle loss and atrophy

suggests that the morphogenic or regenerate-forming activities are favored at the expense of the wound-healing activities. This is the opposite of the situation encountered with muscle-gland incubates. Referring back to the process occurring in vitro, it is evident that as the leg stump begins to recover after leg removal, there is a large amount of damaged muscle tissue and very little regenerative tissue. The need to close the wound results in wound-healing activity with the formation of epithelial sheets and pigment deposition. Later on, the muscle tissue is all but gone and large amounts of regenerate tissue are present and regenerate formation is very active. Thus, it might be anticipated that the presence of muscle tissue would encourage a wound-healing response while the presence of additional regenerate tissue would encourage regenerate-forming responses. At the same time, incubates of neither tissue appeared to have any significant effect on the output of the prothoracic gland.

CONCLUSIONS

The extrapolation of in vitro findings to the processes occurring in vivo is always hazardous. This is especially true when the in vivo processes themselves are only partially understood. On the other hand, the in vitro method described has certain advantages over the in vivo methods presently used. One of these advantages is precise control over the immediate chemical and physical environment of the tissue explants being tested. A second advantage is that the living tissues can be observed repeatedly at high magnification, permitting "before and after" studies on a single specimen at the cellular level. A third ad-

123

vantage is that all the tissues except those being tested can be excluded from consideration. This permits a precise analysis of the interactions occurring between individual glands and between glands and target tissues without interference from substances of unknown origin or from other tissues, such as the ventral nerve chain, which cannot be removed without killing the test animal.

These advantages, then, sufficiently make up for the difficulties involved in extrapolating from in vitro to in vivo conditions and lead to some cautious conclusions. The first of these conclusions is that prothoracic gland incubates contain diffusible substances that produce visible effects on the regenerating leg of a nymphal cockroach.

The second conclusion is that the amount of the diffusible substance that is produced by the prothoracic gland in vitro can be controlled either by increasing or decreasing the number of prothoracic glands incubated or by permitting the prothoracic gland to interact with other endocrine tissues during the incubation period.

The third conclusion is that the interactions of various secretory tissues with the prothoracic glands in vitro may be stimulatory, as with the prothoracic ganglion and the corpus allatum-cardiacum complex, or inhibitory, as with the brain.

The fourth conclusion is that other nonendocrine tissues do not show significant stimulatory or inhibitory interactions with the prothoracic gland in vitro.

Attractive as speculation on these findings may appear, there is, as yet, no direct evidence for the presence of hormones per se in the gland incubates in question. It would seem worthwhile, however, to give further consideration to the implications of these experiments.

The first implication is that the substance being produced by the prothoracic glands is closely related to the classic molting hormone, ecdyson. Preliminary efforts in this laboratory to identify this substance have been unsuccessful. Two approaches are being used. The first of these is the injection of minute amounts of 20-hydroxy ecdyson[1] into chambers containing leg regenerates. While effects similar to those produced by the incubates of two pairs of prothoracic glands have been obtained, it appears doubtful that the substance contained in the prothoracic gland incubate is identical with the purified hormone. Efforts to extract the active substance from the incubates and to submit it to bioassay have also proved inconclusive.

The second implication is that there may be a diffusible substance secreted by the brain, which, in effect, participates in regulating the secretory activity of the prothoracic gland by decreasing its output. Presumably, this substance would function as an antagonist to the stimulatory effects of the classical brain hormone or of the corpora allata, or both. The suggestion that such a substance may occur in cockroaches was originally made by Gersch (1962) who reported two substances with endocrine activity in the brain; one of these acted as a prothoracic gland stimulator and the other as its physiological antagonist. Recently, Langley (1967) suggested that under certain conditions the brain acted as an inhibitory center and delayed the formation of the puparium of the tsetse fly.

While these findings are suggestive, no conclusive evidence has been produced to date, and the existence of an inhibitory brain hormone that may be part of a feedback mechanism controlling the activity of the prothoracic glands remains hypothetical.

REFERENCES

BODENSTEIN, B. (1955). Contributions to the problem of regeneration in insects. J. Exp. Zool. 129, 209–244.

DAY, M. F. (1951). Wound healing in the gut of the cockroach, Periplaneta. Australian J. Sci. Ser. B. 5, 282–284.

GERSCH, M. (1962). The activation hormone of the metamorphosis of insects. Gen. Comp. Endocrinol. Suppl. 1, 322–329.

GILBERT, L., AND SCHNEIDERMAN, H. (1959). Pro-

[1] The 20-hydroxy ecdyson was supplied by John N. Kaplanis of the Insect Physiology Pioneering Research Laboratory, United States Department of Agriculture, Agricultural Research Service, Entomology Research Division, Beltsville, Maryland 20705.

thoracic gland stimulation by juvenile hormone extracts of insects. *Nature* **184**, 171–173.

ICHIKAWA, M., AND NISHITSUTSUJI-UWO, J. (1959). Studies on the role of the corpus allatum in the Eri-silkworm, *Philosamia cynthia* Ricini. *Biol. Bull.* **116**, 88.

LANGLEY, P. A. (1967). Effect of ligaturing on puparium formation in the larva of the tsetse fly, *Glossina morsitans* Westwood. *Nature* **214**, 389–390.

LARSEN, W. (1967). Growth in an insect organ culture. *J. Insect Physiol.* **13**, 613–619.

MARKS, E. P., AND REINECKE, J. P. (1964). Regenerating tissues from the cockroach leg. A system for studying in vitro. *Science* **143**, 961–963.

MARKS, E. P., AND REINECKE, J. P. (1965a). Regenerating tissues from the cockroach, *Leuco-phaea maderae:* Effects of endocrine glands in vitro. *Gen. Comp. Endocrinol.* **5**, 241–247.

MARKS, E. P., AND REINECKE, J. P. (1965b). Regenerating tissues from the cockroach leg: Nutrient media for maintenance in vitro. *J. Kansas Entomol. Soc.* **38**(2), 179–182.

MCDANIEL, C. N., AND BERRY, S. J. (1967). Activation of the prothoracic glands of *Antheraea polyphemus. Nature* **214**(5092), 1032–1034.

O'FARRELL, A., AND STOCK, A. (1953). Regeneration and the molting cycle in *Blattella germanica* L. I. Single regeneration initiated during the first instar. *Australian J. Biol. Sci.* **6**(3), 485–500.

SCHARRER, B. (1948). The prothoracic glands of *Leucophaea maderae. Biol. Bull.* **95**, 186–198.

WIGGLESWORTH, V. B. (1959). The action of growth hormones in insects. *Symp. Soc. Exp. Biol.* **11**, 203–227.

Cockroach Leg Regeneration: Effects of Ecdysterone in vitro

E. P. MARKS
R. A. LEOPOLD

Although the appearance of cuticle-like deposits on cultures of insect epidermal tissues have been reported (1–6), there has been considerable confusion over the nature of the materials being deposited (2–6), their relation with the epithelial cells (3, 6), and the conditions under which they are deposited (1, 4). The chemical nature of the deposits (1–5) and the hormonal requirements of explants (2, 3, 6) remain largely undefined. To elucidate some of these problems, we have made a series of parallel studies in vivo and in vitro using regenerating leg tissues from the cockroach *Leucophaea maderae.*

The mesothoracic legs of late instar nymphs were removed at the trochantero-femoral articulation 24 hours after ecdysis. At specific times, 8 to 42 days after removal of the leg, the coxae containing the regenerating legs were removed from the insects and the regenerates were dissected. Some of these regenerates were fixed, sectioned, and subjected to histochemical examination, while others were placed in Rose multipurpose tissue chambers under dialysis strips and cultured in M-18 (7) nutrient medium. The histochemical procedures for investigating the cuticular sheaths included the triple stain technique for demonstration of protein and carbohydrate (8), the thiazine red dichroic method for determination of birefringence (9), and the fluorescent chitinase test for chitin (10). In conjunction with the triple stain technique, several prior treatments were used to verify the specificity of the staining reaction. These included acetylation, deacetylation, and pepsin digestion (11). When dissected 25 days after leg removal, the regenerates were covered with a thin gelatinous sheath that showed a positive PAS (periodic acid–Schiff) reaction (Figs. 1 and 2). Acetylation prevented this reaction while deacetylation restored it, confirming that the PAS-positive reaction was the result of the presence of 1,2-glycols contained in carbohydrate substances. Prior treatment with pepsin also prevented the PAS reaction. These results indicate that the sheath consists of carbohydrate complexes conjugated with protein, possibly a glyco- or mucoprotein which dissociates upon pepsin hydrolysis. The sheath displayed neither dichroic birefringence nor affinity for the fluorescent-chitinase conjugate, an indication that chitin was not present.

When the regenerates were allowed to develop for 40 to 42 days before dissection, they were found, upon removal, to be covered with a wrinkled membrane bearing spines and small setae (Fig. 3) that had developed between the original sheath and epithelial cells. This membrane was PAS-negative, displayed dichroic birefringence, and showed affinity for the fluorescent-chitinase conjugate (Fig. 4). This layer, which had replaced the initial sheath, thus represented the early stage of the developing chitinous cuticle.

When 25-day-old leg regenerates were placed in culture, epithelial cells quickly migrated over the injured areas and a new sheath was deposited over the surface (2). Some of the epithelial cells developed hairlike processes that extended out into the gelatinous sheath, but in no case did a recognizable cuti-

cle develop. These activities continued for as long as 30 days.

To investigate the hormonal induction of cuticle formation in vitro, Rose chambers were set up in pairs so that each member of the pair contained a leg from opposite sides of the insect. One chamber of the pair was treated with ecdysterone (12), while the other was treated either with water or with another isomer of ecdysone, 22-isoecdysone, which has been reported to be inactive in the Calliphora assay system (13, 14). When such chambers were treated 4 days after explantation with water solutions of ecdysterone (2.5 to 12 μg per milliliter of medium), evidence of deposition of cuticle was apparent by the end of 7 days. A thin refractile layer appeared between the epithelial cells and the gelatinous sheath; the epithelial cells began to round out and became filled with a granular material; and, in a few days the refractile layer thickened and became wrinkled (Fig. 5). At the same time, the hairlike epithelial cells began to thicken, and at the base a socket began to form. By day 14 after treatment, the cuticle that showed affinity for the fluorescent-chitinase conjugate began to darken, forming a rugose covering over the explant (Fig. 6 and Table 1).

All the 14 regenerates treated with ecdysterone developed a recognizable cuticle within 14 days, and 8 devel-

Table 1. The effect of ecdysone analogs on the deposition of cuticle in vitro in 25-day-old cockroach leg regenerates.

Treatment	Dose (μg/ml)	Number of regenerates		
		Treated	Developing cuticle	Developing setae
Water	10*	4	0	0
22-Isoecdysone	12	10	0	0
Ecdysterone	12	4	4	3
Ecdysterone	5	6	6	2
Ecdysterone	2.5	4	4	3

* In microliters.

127

Fig. 1. Sheath (a) deposited by epithelial cells (b) after 25 days in vitro.   Fig. 2. PAS-positive sheath (a) formed by epithelial cells (b) of 30-day-old leg regenerate in vivo. Basement membrane (f) also yielded a PAS-positive reaction (triple stain).   Fig. 3. Surface of leg regenerate (44 days in vivo) just before molting shows sheath (a) and chitinous cuticle with setae (c) and rugose surface (e).   Fig. 4. Fluorescent chitinase test indicates presence of chitin in both the cuticle (d) and seta (c) in leg regenerate just before molting.   Fig. 5. Cuticular deposit (d) appears in vitro between sheath (a) and epithelial cells (b) in leg regenerate treated with β-ecdysone. Figure 1 shows paired control treated with water.   Fig. 6. Seta (c), cuticular deposits (d), and rugose surface (e) formed in vitro are evident 27 days after treatment with β-ecdysone. Cuticle is pigmented. [Scale, 50 μ]

128

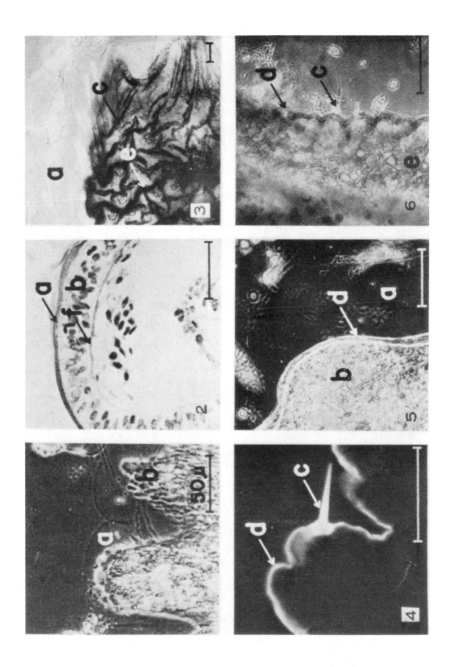

129

oped setae. Doses as low as 2.5 μg per milliliter of medium were 100 percent effective. None of the 14 control regenerates produced cuticle or setae, and there was no apparent difference between those treated with water and those treated with 22-isoecdysone.

At the end of the 14-day period, all control chambers were washed out, filled with fresh medium, and treated with ecdysterone. After eight more days, all the original water control explants showed recognizable cuticle. Those that had been treated with 22-isoecdysone also developed cuticle but required two to five more days to do so.

The chemical composition and function of the gelatinous sheath are only partially known. We found similar sheaths in cultures of epidermal tissues from embryos of the tobacco hornworm *Manduca sexta* (Johannson) and the differential grasshopper *Melanoplus differentialis* (Thomas). Because these deposits occur only on the outer surface of epidermal tissues, they may serve to protect the delicate epithelium and developing setae, while the underlying cuticle is secreted.

The appearance of the chitinous cuticle was unmistakable because the setae and surface sculpturing were similar to that found at the time of ecdysis on regenerates that develop in vivo. The cuticle formed in vitro, however, remained thin and only partially developed. The heavy layer of endocuticle that is formed in vivo at the time of ecdysis was largely absent, a situation similar to that reported by Miciarelli *et al.* (4).

With the possible exception of Sengel and Mandaron (5), none of the earlier workers succeeded in inducing the deposition of cuticle by explants of epidermal tissue in vitro. Sengel and Mandaron obtained what seemed to be cuticular deposits on leg imaginal disks of late instar *Drosophila* larvae by treating them with ecdysone. Unfortunately, both brain and ring gland tissues were present in these cultures, and the results obtained were such that the source

of stimulation and the nature of the cuticular deposits remain uncertain. Miciarelli *et al.* (4) and Demal (1) used explants of epidermal tissues from immature insects late in their developmental cycle, and cuticle deposition was spontaneous. Larsen (3) and Ritter and Bray (6) used long-term cultures of tissues of mesodermal origin. Larsen obtained spontaneous deposits of unknown composition, and Ritter and Bray found "chitin"-containing crystals that were spontaneously deposited on the glass surface in cultures of blood clots.

In our study, the secretory activities of the epidermal cells have been changed from the deposition of sheath material to that of chitinous cuticle. This change, which can be induced by ecdysterone in low doses, is not induced by much larger doses of an inactive ecdysone analog. This system, in which the epidermal cells are available for microscopic examination throughout both secretory processes, provides an excellent means of studying the effects of hormones on the process of cuticle deposition.

References and Notes

1. J. Demal, *Ann. Sci. Natur. Zool. Biol. Anim.* **18**, 155 (1956).
2. E. P. Marks and J. P. Reinecke, *Science* **142**, 961 (1964).
3. W. Larsen, *J. Insect Physiol.* **13**, 613 (1966).
4. A. Miciarelli, G. Sbrenna, G. Colombo, *Experientia* **23**, 64 (1967).
5. P. Sengel and P. Mandaron, *C. R. Hebd. Seances Acad. Sci. Paris* **268**, 405 (1969).
6. H. Ritter and M. Bray, *J. Insect Physiol.* **14**, 361 (1968).
7. Available from Grand Island Biological Co., Grand Island, N.Y. 14072.
8. M. Himes and L. Moriber, *Stain Technol.* **31**, 67 (1956).
9. H. Fuller, *Zool. Anz.* **174**, 125 (1965).
10. M. A. Benjaminson, *Stain Technol.* **44**, 27 (1969).
11. A. G. E. Pearse, in *Histochemistry: Theoretical and Applied* (Little, Brown, Boston, ed. 2, 1961), p. 998.
12. Available from Mann Research Laboratory, New York 10006.
13. A. Furlenmeier, A. Furst, A. Langemann, G. Waldvogel, P. Hocks, V. Kerb, R. Weichert, *Helv. Chim. Acta* **50**, 2387 (1967).
14. Provided by Dr. John Siddall, Zeocon Corp., Palo Alto, Calif. 94304.
15. Supported in part by a grant from the Kales Foundation.

EFFECTS OF ECDYSTERONE ON THE DEPOSITION OF COCKROACH CUTICLE *IN VITRO*

E. P. MARKS

The production of cuticular materials by insect tissue *in vitro* in response to stimulation by the molting hormone was discussed in a recent review (Marks, 1970). In these early studies, relatively large amounts (3 to 25 μg/ml of medium) of the hormone were used, the period of exposure ranged from 2 to 10 days, and no systematic attempt was made to determine the effects of different doses. In a recent study, Marks and Leopold (1971) reported that cuticle deposition in cockroach leg regenerates could be induced *in vitro*. High doses (2 to 25 μg/ml) given over long periods (7 to 10 days) resulted in the deposition of a heavy cuticle complete with well-defined setae; smaller doses produced a thinner cuticle on which droplets of chitin-bearing material accumulated. The lowest concentration that induced any deposition was 0.1 μg/ml given over a period of 7 days. These results suggested that a quantitative study of hormonal induction of cuticular deposition *in vitro* might facilitate the investigation of the mechanisms whereby ecdysterone activates the epidermal cells and of the process of cuticle deposition.

In the present study, the effects of three dose-related variables were investigated: the concentration of the hormone, the duration of exposure, and interruption of the dose.

METHODS

Late-instar nymphs of the Madeira cockroach, *Leucophaea maderae* (F.) were removed from the laboratory colony immediately after molting. The mesothoracic legs were removed 24 hr later at the femora-trochenteral joint. The operated insects were held for 24–26 days at room temperature in paper cups and fed dog food and water. At the end of this period, the coxae were dissected, and the regenerating legs, which at this stage comprised about 2 mm^3 of tissue each, were removed and placed in multipurpose tissue chambers (Rose, 1954) filled with 2 ml of M20 nutrient medium containing 5% fetal bovine serum. These were maintained at 27° C. The ecdysterone dissolved in water (1.0 or 0.1 μg/μl) was held under refrigeration, and a new solution was made up every 30 days. The hormone was injected into the chambers, and at the end of the period of exposure, the treated medium was removed and the chambers were rinsed twice and refilled with fresh medium. All chambers were observed periodically for 14 days after exposure to the hormone and then scored for the presence of cuticle. Since the thickness of the cuticle varied with the dose, a positive score was given only if both chitin droplets and a cuticular membrane were present on at least two areas of the explant (Marks and Leopold, 1971).

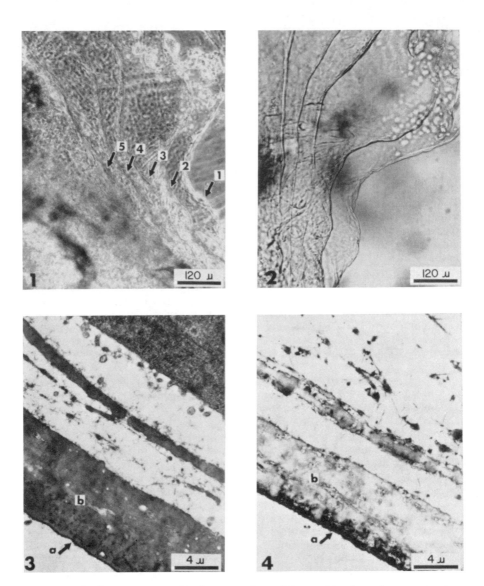

FIGURE 1. Leg regenerate with five successive cuticles resulting from five separate doses of ecdysterone. Secretion of most recent cuticle is still incomplete, and visualization is obscured by tissue.

FIGURE 2. Same specimen as Figure 1 after epidermis has retracted. Most recent cuticle is complete, and five cuticles are clearly visible.

FIGURE 3. Electron micrograph of inner of two cuticles secreted by a single leg regenerate. Surface of epidermal cell is visible at upper right. Epicuticle (a) is clearly distinguishable from less dense procuticle (b).

TABLE I

Comparison of responses obtained by administering a 1-μgd dose of
ecdysterone in different ways

Number of exposures	Length of exposure (days)	Dose (μg/ml)	Total dosage (μgd)	Total days		Frequency of response (%)
				Exposed	Elapsed*	
1	1	1.00	1	1	1	16
2	1	0.50	1	2	5	25
4	1	0.25	1	4	13	20

* Includes time between repeated doses.

Two types of dosage were investigated, those in which the period of exposure
was constant and the concentration of the hormone was varied and those in which
the concentration was constant and the exposure was varied. Also, two types of
exposure were used, continuous and interrupted. For the interrupted exposure,
the hormone was injected for a given period and then the chambers were rinsed,
refilled, and held for 3 days before the next exposure; this procedure was repeated
until the total dose was given. The total dose received by an explant was expressed as the number of days of exposure multiplied by the concentration of the
hormone in μg/ml of medium. Thus, the expression of dosage in microgram
days (μgd) was common to all the experiments.

RESULTS

Since time-dose studies often involve long exposures (or as with interrupted
doses, long periods during which the total dosage is delivered), it was necessary
to establish that an effective dose of the hormone was maintained under the conditions of the experiment and that the explant retained its sensitivity to the hormone
over the entire period of exposure. Therefore, three chambers, each containing
two explants, were injected with ecdysterone to give a final concentration of 2.5
μg/ml and incubated for 30 days. At the end of this period, all the explants had
well-developed cuticle. The medium was then removed from these chambers and
injected into other chambers containing fresh explants. Cuticle appeared on all the
new explants within 10 days. In another experiment, ecdysterone (2.5 μg/ml)
was injected into a chamber containing two explants. The medium from this
chamber was transferred after 7 days to another containing two more fresh
explants and transferred to a third chamber after 7 more days. Then after 7
days in the third chamber, the medium was returned to the first chamber for 7
days (no new chambers were available) and finally on again to the second one.
Within 10 days after the first dose, the explants in all three chambers showed
evidence of cuticular deposition. Furthermore, when the explants were exposed
to the medium the second time, they developed a second cuticle inside the first.
Moreover, when a chamber containing fresh explants was injected with the aged

FIGURE 4. Electron micrograph of outer cuticle of same specimen as in Figure 3. Again,
epicuticle (a) can be distinguished from procuticle (b). Similarity in the structure of the two
cuticles is apparent.

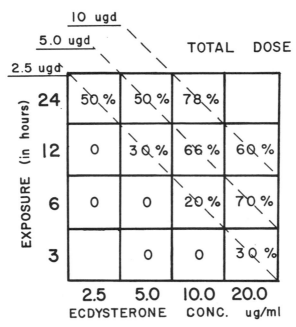

FIGURE 5. Diagram showing relative effect of concentration of hormone, time of exposure, and total dose (μgd) received on the frequency (%) of cuticle deposition by cockroach leg regenerates treated *in vitro* with ecdysterone.

medium some 35 days after the start of the experiment, cuticle appeared within 10 days. Superficially at least, this cuticle was no different from the first one in the series. The same experiment was repeated with a lower total dose (0.5 μg/ml for 5 days). After passage through three chambers, enough ecdysterone still remained in this medium to induce cuticular deposition on fresh explants in a fourth chamber.

Apparently neither metabolism by as many as 10 regenerates nor exposure of the hormone to breakdown over a period of 30 days was sufficient to reduce the dosages to a level below the threshhold. Marks and Leopold (1971) demonstrated that untreated regenerates used as solvent (H_2O) controls could be induced to respond to ecdysone treatment, even after 14 days *in vitro*. This ability to respond over a period of time was further explored by treating a series of chambers with ecdysterone (2.5 μg/ml). After five days, the treated medium was removed, and the chambers were rinsed and refed. As soon as cuticular deposits appeared on the explants, the chambers were refed and retreated with ecdysterone. This cycle was repeated until the tissues failed to respond. In this way, we were able to obtain as many as five separate cuticles from a single explant (Figs. 1 and 2), and the tissues apparently retained the ability to respond to treatment over 50 days.

One explant that had produced two cuticles was examined by electron microscopy: A section of the inner (second) cuticle (Fig. 3) was compared with a

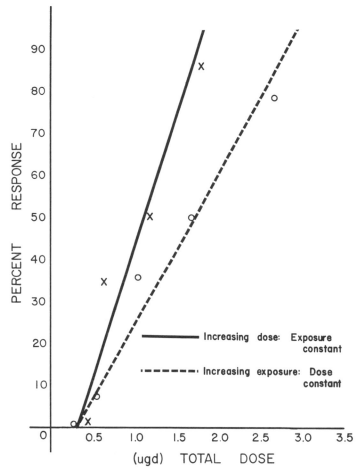

FIGURE 6. Graph showing the effect of the mode of delivery of a given total dose of ecdysterone on the frequency of cuticle deposition in cockroach leg regenerates.

section of the outer (first) cuticle (Fig. 4). Except for their position relative to the explant, it was difficult to distinguish between them. Each cuticle had an epicuticle (a) and a procuticle (b), evidence that they were the products of separate secretory cycles.

Once the long-term stability of the experimental system was established, a series of time-dosage tests was made. In the first series, high concentrations of ecdysterone (2.5 to 20 μg/ml) and short exposures (3 to 24 hr) were used. The minimum number of replicates for each determination was 10, and the results are given in Figure 5. The response to the hormone could be increased by increasing either the concentration or the length of exposure and was reasonably consistent in

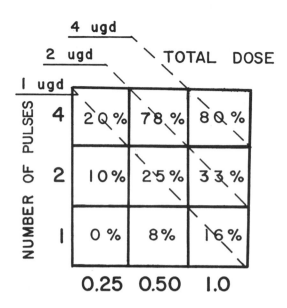

FIGURE 7. Diagram showing the relative effect of hormone concentration, number of exposures, and total dose received on the frequency (%) of cuticle deposition by cockroach leg regenerates treated *in vitro* with ecdysterone.

terms of total dose (μgd). When the exposure was 12 hr or less, the 50% response level lay between 2.5 and 5 μgd. At shorter intervals, the ecdysterone concentration required became so high, further experiments were impractical.

In the second series, the concentration of ecdysterone was lowered, and the period of exposure was increased. In one experiment, the period of exposure was held constant at 7 days, and the concentrations of ecdysterone varied from 0.05 to 0.50 μg/ml. In another experiment, the concentration was held constant at 0.50 μg/ml, and the periods of exposure varied from 12 hr to 7 days. The minimum number of replicates used to obtain each point was 10, and the results are given in Figure 6. The data show that time of exposure and concentration of hormone make roughly equal contributions to the effectiveness of the total dose. The dose required to give a 50% response lay between 1.0 μgd with a 7-day exposure and 1.4 μgd with a 3-day exposure. This was roughly half the dosage required for the short-term experiments.

Since the preliminary experiments showed that repeated large doses (10 μgd or greater) produced a series of separate cuticles, an additional series of experiments was made with repeated exposures of 1 day and concentrations of 0.25, 0.50, and 1.0 μg/ml. The minimum number of replicates was 12. After each dose, the chambers were washed with 3 changes of nutrient and held 3 days before the dose was repeated; as many as 4 such doses were given over a period of 13 days. The results are summarized in Figure 7. In no case did a recognizable cuticle appear until after the last dose was given. Apparently, the effects of the repeated doses are cumulative, and are independent of the dose given in a single exposure and of the time elapsed between the first and last exposure (Table I).

DISCUSSION

Oberlander (1969a), working with wing discs of *Galleria*, reported that ecdysone remained active in cultures for as long as three days and found no evidence that the tissues metabolized appreciable quantities of the hormone. His findings and the results of the present experiments stand in sharp contrast to those of Shaaya (1969), Karlson and Bode (1969), and Ohtaki, Milkman, and Williams (1968), all of whom found that the half-life of ecdysone analogs in living insects was at best a few hours. The most obvious explanation for this difference between the *in vivo* and *in vitro* results is that the hormone is either rapidly inactivated or excreted by living insects; neither would occur to any great extent in an *in vitro* system. However, a recent study by Richman and Oberlander (1971) indicates that the situation may be more complex than this.

The fact that multiple doses of hormone will stimulate the production of multiple cuticles sheds some light on questions raised earlier by Marks and Leopold (1971) as to why the *in vitro* system produced cuticles containing only small amounts of procuticle. Since the cuticle produced by a second dose of hormone was essentially similar to the one laid down 12 to 15 days earlier, the mechanisms producing the cuticular materials must have retained the ability to operate *in vitro* over considerable periods. Moreover, since the explant was able to secrete as many as 5 similar cuticles, the incompleteness of the first one could not have occurred because of the inability of the explant to produce enough material. What occurred is apparently the normal response to ecdysterone stimulation, and the completion of cuticular deposition *in vivo* probably requires either an additional stimulus of a different type or another source of material. In any case, what occurred *in vitro* is only the first of a series of steps that leads to the formation of a complete cuticle, and this first step was repeated with each dose.

Oberlander (1969a, 1969b) found that a 2-hr exposure to ecdysone at a concentration of 3 μg/ml was sufficient to induce morphogenic changes in the wing discs of *Galleria*. In the present study, I found that the overall (7-day) threshold for cuticular deposition was 0.35 μgd, close to the 0.25 μgd required to induce morphogenesis in Oberlander's studies; with a 3-hr exposure, it was between 1.25 and 2.5 μgd. Thus, as the exposure decreased, the total dosage required to induce deposition increased, indicating that this system is somewhat less sensitive than Oberlander's. This lowering of the threshold with increasing exposures suggested that the accumulation of some substance or physiological event was taking place. Evidence that such a buildup did indeed occur was obtained from the experiments made with discontinuous doses. The subthreshold doses given over long periods eventually triggered a response; furthermore, when the total dose was the same, about the same response was obtained, whether the dosage was spread over 1, 5, or 13 days.

Since each chamber was washed three times and held for three days between exposures, it is difficult to believe that a quantitative buildup of the ecdysterone

137

occurred unless it was tightly bound by the tissue. Thus, the relationship between concentration and exposure and the quantitative accumulation of interrupted doses suggested that it was the "event" rather than a substance that accumulated. Like the "events" resulting from exposure to ionizing radiation, these are apparently discrete and of an all or none nature.

Ohtaki, Milkman, and Williams (1968), working with *Sarcophaga*, assayed the concentration of ecdysone present at various stages of development and found that it never reached the level required to induce pupation in an isolated abdomen, but it was never entirely absent. They concluded that the presence of a low titer of hormone over a period of time resulted in the accumulation of a series of covert events that accumulated until they finally produced an overt response. The minimum period for the process of accumulation, regardless of the concentration of hormone, was 8.5 hr, which is in the same range as the 3 to 12 hr reported by Oberlander (1969a, 1969b) and the 6 and 12 hr required for a 50% response in my experiments. These results give strong support to the hypothesis of Ohtaki, Milkman, and Williams (1968). Such an all or none event might be either the tenacious occupation by the ecdysterone molecule of a receptor site on or within the cell or the passage into the cell of a single molecule of the "macromolecular factor" proposed by Williams and Kambysellis (1969). In the latter case, the ecdysterone would act to mediate the passage of the molecule through the cell membrane.

Since the entire series of events leading to cuticle deposition can be induced in the same explant as many as 5 times, it is apparent that once deposition has been induced and the cells have discharged their cuticular materials, they also discharge their accumulated events and are ready to start a new cycle, again responding in the same way as untreated tissue.

The author acknowledges his indebtedness to J. G. Riemann for the electron micrographs in Figures 3 and 4. This work was supported in part by a grant from the Kales Foundation.

LITERATURE CITED

KARLSON, P., AND C. BODE, 1969. Die Inactivierung des Ecdysons bei der Schmeissfliege *Calliphora erythrocephala* Meigen. *J. Insect Physiol.,* **15**: 111–118.

MARKS, E. P., 1970. The action of hormones in insect cell and organ cultures. *Gen. Comp. Endrocrinol.,* **15**: 289–302.

MARKS, E. P., AND R. A. LEOPOLD, 1971. Deposition of cuticular substance *in vitro* by leg regenerates from the cockroach, *Leucophaea maderae* (F.). *Biol. Bull.,* **140** : 73–83.

OBERLANDER, H., 1969a. Effects of ecdysone, ecdysterone, and inokosterone on the *in vitro* initiation of metamorphosis of wing discs of *Galleria melonella*. *J. Insect Physiol.,* **15** : 297–304.

OBERLANDER, H., 1969b. Ecdysone and DNA synthesis in cultured wing discs of the greater wax moth, *Galleria mellonella*. *J. Insect Physiol.,* **15** : 1803–1806.

OHTAKI, T., R. MILKMAN AND C. WILLIAMS, 1968. Dynamics of ecdysone secretion and action in the flesh fly, *Sarcophaga peregrina*. *Biol. Bull.,* **135** : 322–334.

RICHMAN, K., AND H. OBERLANDER, 1971. Effects of fat body on α-ecdystone induced morphogenesis in cultured wing discs of wax moth, *Galleria mellonella*. *J. Insect Physiol.,* **17** : 269–277.

ROSE, G., 1954. A separable and multipurpose tissue culture chamber. *Texas Rep. Biol. Med.,* **12** : 1074–1083.

SHAAYA, E., 1969. Untersuchungen über die verteilung des Ecdysone in verschiedenen Geweben von *Calliphora erythrocephala* und über sein biologische Halbwertszeit. *Z. Naturforsch.* **24** : 718–721.

WILLIAMS, C. M., AND M. KAMBYSELLIS, 1969. *In vitro* action of ecdysone. *Proc. Nat. Acad. Sci., U. S.,* **63** : 231.

Regeneration in Echinoderms and Urochordates

Regeneration in the Sea Cucumber *Leptosynapta*
I. THE PROCESS OF REGENERATION

GERALD N. SMITH, JR.

Holothurians of the genus *Leptosynapta* respond to environmental stress by casting off portions of the body column (Clark, 1899; Pearse, '09). Regeneration of the complete animal from the most anterior piece has been assumed since Clark (1899) reported that any piece of *Leptosynapta* which includes both the mouth and part of the adjacent digestive tract survives and apparently regenerates. No other piece survives. Although Clark reported perceptible growth of anterior pieces two weeks after section, no detailed data on the process of regeneration exists.

This report describes the process and time course of regeneration of the missing posterior part by anterior pieces of *Leptosynapta crassipatina*. Anterior parts include—in addition to the mouth and part

of the adjacent digestive tract—the nerve ring, the calcareous ring, and the entire water vascular system. These parts are studied histologically for evidence of participation in the regeneration process. Evidence for a growth region in this area is sought.

MATERIALS AND METHODS

Specimens of *Leptosynapta crassipatina* collected on inter-tidal sand flats near the Florida State University Marine Laboratory at Turkey Point in Franklin County, Florida, were used in all experiments. Both Clark and Pearse worked with the common synaptid of New England which they called L. *inhaerens* (O. F. Muller, 1738). This is still common practice (Smith, '64), but Heding ('28) concluded, after comparing ossicles and ciliated funnels of L. *inhaerens* from Naples, Italy, with those from L. *inhaerens* from Woods Hole, Massachusetts, that the New England variety is a separate species. He chose the next available name, L. *tenuis* (Ayers, 1851). Accordingly, I will use L. *tenuis* when referring to this species. L. *crassipatina* is the synaptid of the eastern Gulf of Mexico. It can be distinguished from L. *tenuis* only by microscopic examination of the ossicles (Clark, '24). The experiments were begun within three days of collection. Animals were maintained in glass aquaria with artificial sea water (Harvey, '45) at 21–23 °C. The water was changed daily, and observations were made at either 12 or 24 hour intervals. This procedure was satisfactory for short experiments, but for very long experiments, and to control for the effects of starvation, animals were kept in running sea water with a supply of mud and detritus on sea tables at the marine laboratory. Experimental animals, as well as controls, were relaxed in 0.54 M $MgSO_4$ for 20 to 30 minutes before each operation. The operations were performed on a paraffin backing with a razor blade.

Animals intended for histological study were relaxed in $MgSO_4$ for at least one hour and then simultaneously killed and fixed by immersion in fixative. Bouin's fluid, modified for decalcification by substitution of formic acid for acetic acid (Lillie, '65) was used routinely. After 48 hours of fixation, decalcification of both spicules and clacareous plates was complete. Paraffin sections were prepared at 10 μ. Most sections were stained in Masson's Trichrome connective tissue stain. To demonstrate mitotic figures, either hematoxylin with a fast green counterstain (Humason, '67), or Spicer's Feulgen procedure (Lillie, '65) was used. Dilute toluidine blue showed connective tissue changes.

A modification (Hale, '64) or Vogt's ('25) vital staining technique was used to provide extra markers in the relatively featureless posterior body column. Squares of cellophane (4 cm) were stained in 1% solutions of either Nile blue or neutral red. These pieces were then dried and cut into 2 mm strips. The strips were placed on the body wall of relaxed animals for 15 to 30 minutes. Only those parts in contact with the cellophane are stained. The stain does not diffuse significantly to other cells (Vogt, '25).

RESULTS

Morphology

In external appearance, *Leptosynapta* is a whitish tube with a ring of 12 pinnate tentacles at the anterior end. The tentacles are 2–4 mm in length and bear three to five pairs of digits. Length and thickness of the body column vary with the tone of the body wall musculature. Length of a typical relaxed adult specimen is 150 mm; diameter is from 4–6 mm. Calcareous inclusions and most internal organs can be seen through the translucent body wall. The longitudinal muscles of the body wall appear as white bands, marking the five ambulacra that define the pentamerous symmetry of the organism. By convention, the mid-ventral ambulacrum is labelled radius A; the other are designated B, C, D, and E around the circumference in a clockwise direction from A, looking along the animal from mouth to anus. Two tentacles arise from dorsal interradius CD, and from each of the two ventral interradii, AB and EA. Three tentacles each arise from interradii BC and DE.

The water vascular system in *Leptosynapta* is composed of the water ring, the stone canal, the polian vesicle, and the tentacular ampullae, as well as the oral tentacles which are modified tube feet. The bases of the tentacles enfold the calcareous ring at the junctions of the plates, such that each tentacle base enfolds half of each of two plates. The plates protude, by approximately half their height, into the central canal of the tentacle, dividing the lower part of the canal into an outer and inner compartment. Each tentacle is supplied with an ampulla which leads into the inner compartment through a semi-lunar valve (fig. 1). The ampullae connect the tentacles to the anterior surface of the water

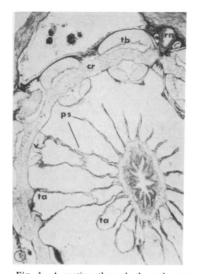
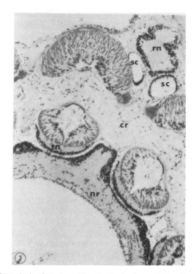

Fig. 1 A section through the calcareous ring (cr) showing the tentacle base (tb), the tentacular ampullae (ta) and the valve (v) through which the tentacular ampulla and the tentacular canal are connected. The pharyngeal suspensors (ps) show clearly. Note also the radial nerve (rn) and the statocysts (sc). Toluidine blue. × 100.

Fig. 2 A section through the nerve ring (nr), calcareous ring (cr), statocysts (sc), and radial nerve (rn). Note also the tentacle base (tb) and the tentacular nerve (tn)' Trichrome. × 250.

ring which encircles the pharynx two mm posterior to the calcareous ring.

The water ring has two additional appendages. From the ventral side a single polian vesicle extends posteriorly for 1.5 cm. The stone canal arises from the dorsal side of the water ring and runs anteriad in the anterior margin of the dorsal mesentery. The stone canal opens into the coelom, through one or more madreporites.

The integrity of the oral complex as a morphological unit is reinforced by the pharyngeal suspensors. These are sheets of connective and muscle tissue radiating from the pharynx to the plates. Each pair of suspensors enfolds one tentacular ampulla throughout its length. From the level of the ampullar valve to the posterior margin of the plate, a suspensor inserts on the plate to either side of the ampulla. The other end of each suspensor is attached to the pharynx (fig. 1). Posterior to the plates, the suspensors encircle the ampulla and form a U-shaped sheath. The ampulla is in the cup of the U, and the arms of the U attach to the pharynx.

The mouth is an oval opening in the buccal membrane in the center of the ring of tentacles. The pharyngeal section of the digestive tract runs from the mouth to the level of the water ring. It is characterized by a thick layer of circular muscle. The nerve ring, encircling the pharynx at the anterior margin of the plates, is just posterior to the buccal membrane, and is closely applied to the calcareous ring (fig. 2). Anteriad, 12 tentacular nerves arise from the ring and course along the central surface of the tentacle. At each radius, a nerve passes through a pierced plate and runs posteriorly, within the body wall, to the anus. Each radial nerve is overlain by a longitudinal muscle band which inserts on the outside of the calcareous ring and runs to the anus.

The gut and its suspending mesenteries provide the best linear morphological markers in the relatively featureless body

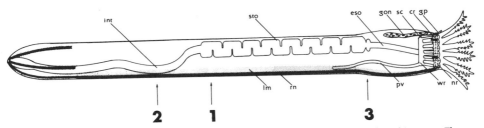

Fig. 3 Diagram of *Leptosynapta* showing the three levels of section discussed in this paper. The anterior piece regenerates the missing posterior. The levels chosen present the organism with three distinctly different problems of regeneration and can be located easily by macroscopic observation. Abbreviations used are int, intestine; sto, stomach; eso, esophagus; gon, gonad; sc, stone canal; cr, calcareous ring; gp, gonopodium; nr, nerve ring; wr, water ring; pv, polian vesicle; rn, radial nerve; lm, longitudinal muscle.

column (fig. 3). There are three externally visible divisions of the gut in *Leptosynapta*, arbitrarily designated esophagus, stomach, and intestine without implying function. Stomach and intestine are about equal in length; the short esophagus constitutes only one tenth the length of the digestive tract.

The esophagus is a thin tube with a smooth external surface. The stomach is yellow and much larger in diameter than the esophagus. Its external surface is highly folded. Finally, the intestine is thinner and quite smooth externally.

The esophagus and stomach are suspended from the midline of interradius CD by the dorsal mesentery. Posteriorly, the intestine is attached to the ventral mesentery along the midventral muscle. The point where the suspension of the gut changes from dorsal to ventral (the stomach-intestine junction) will be called the crossover point for convenience. A third mesentery runs along interradius DE. This lateral mesentery is not attached to the gut, but at the extreme anterior end of the body a mesenteric sheet joins the ventral and lateral mesenteries across muscles A and E.

The haemal system consists of a dorsal or mesenteric vessel and a ventral or free vessel connected through the gut walls by haemal lacunae. The haemal vessels are actually sinuses, since they lack a lining. The lacunae appear to be continuous throughout the length of the gut, forming a connective tissue-fluid complex as de-

scribed by Fish ('67) in *Cucumaria*. The anterior end of the free sinus terminates at the esophagus-stomach junction. At the crossover point, this sinus curves around the gut to maintain its position opposite the suspending mesentery. Posteriorly, the free sinus ends just anterior to the anus. The mesenteric sinus is attached along the junction of the gut and its suspending mesentery throughout its length. At the esophagus-stomach junction, a ring-shaped sinus connects the mesenteric sinus and the free sinus. The haemal fluid is yellow in freshly collected animals or in animals supplied with nutrients, but becomes colorless after a week of starvation conditions.

The junctions of the dorsal and lateral mesenteries with the body wall are lined with rows of ciliated funnels, or urns. These funnel-shaped organs appear to function in the removal of particulate wastes from the coelomic fluid. The funnels aggregate such waste materials and phagocytic amoebocytes to form brown bodies (Hetzel, '65) which are later eliminated from the coelom. Brown body formation is a normal process, but it occurs more frequently during regeneration.

All of the fluid spaces of the body are populated by wandering cells, the coelomocytes. The major types of coelomocytes are the amoebocytes, either petaloid or filiform; the lymphocytes; and the morula cells (Hetzel, '63).

145

The influence of the level of section on the process of regeneration was determined by sectioning animals at one of three levels: (1) just anterior to the crossover point; (2) just posterior to the crossover point; (3) at the esophagus-stomach junction (fig. 3). Changes were noted at 24 hour intervals and are predictable. At the three transection levels utilized in this study the survival percentage approaches 100%, and regenerates appear normal.

The initial event of regeneration is closure of the wound and restoration of coelomic integrity. Regardless of the level of section, strong contraction of the circular musculature closes the wound, and the normal cellular clotting mechanism seals the new coelomic compartment within 30 minutes. After 24 hours, the closure is sufficiently strong to withstand even the large increase in internal pressure generated by vigorous contractions of the body wall musculature. The subsequent events of regeneration are determined by the level of section.

(1) Anterior to the crossover point

When transection is just anterior to the crossover point, the anterior piece lacks the entire intestine. The remaining alimentary canal includes the pharynx, esophagus, and stomach, and is suspended in the dorsal mesentery throughout its length. The ventral mesentery is evident only in the extreme anterior end where it joins with the lateral mesentery to form the mesenteric sheet across muscles A and E.

After 24 hours, the animal is unchanged macroscopically. Between 48 and 72 hours, the stomach begins to change in position and appearance. The highly folded external surface of the stomach becomes smooth, and the gut begins to move from its dorsal attachment to the ventral side of the coelom. This change begins at the posterior end and progresses anteriad. Between 72 and 96 hours the alimentary canal continues to change. The dorsal mesentery recedes, and the ventral mesentery advances, until the normal proportion of the gut attached to

TABLE 1

Time from transection to crossover at Level 1

Time	Experiments	Survivors	Crossovers
hours			
0	20	20	0
36	20	20	0
48	20	20	6
60	20	20	15
72	20	20	19
84	20	20	20

each of these sheets is attained. The esophagus-stomach junction also moves anteriad, until the normal proportion of the two regions is established. After 144 hours of regeneration the animal can no longer be identified as a regenerate by macroscopic examination; it is a complete, though shortened, organism. The stages of regeneration described here are based on several populations of animals and the times reported are idealized. The actual time lapse between section and crossover in a population of 20 animals is shown in table 1. These observations are at 12 hour intervals. With 24 hour intervals the events are predictable.

The mesenteric connections were studied histologically at 24, 48, 72, and 96 hours after operation. Four animals were prepared at each time given, and an additional eight animals were prepared at the 48 hour stage.

After 24 hours, some change is evident. The ventral mesentery appears as a flap along muscle A, and the cell density increases within the connective tissue-fluid complex. There is some degradation of the longitudinal muscles near the wound site, and cellular debris is evident within the coelom. The beginning of the process through which the digestive tract regains its normal position and missing parts is seen by 48 hours. The posterior tip of the digestive tract of the 48 hour animal is attached to the ventral mesentery at the mesenteric haemal vessel. At this level, the lateral and dorsal mesenteries are joined to one another and to muscle D (fig. 4). At more anterior levels, all three mesenteries are joined and attached to the gut at the dorsal haemal vessel (fig. 5). Anterior to this triple attachment, the gut is suspended in the dorsal mesentery. At this level the lateral mesentery is represented only by a row

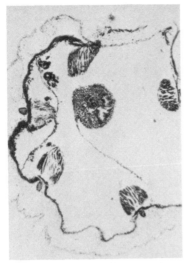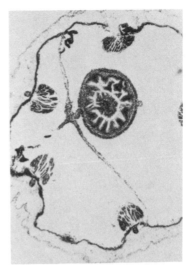

Fig. 4   A section posterior to the new crossover point in an animal 48 hours after transection at level 1. The gut is suspended in the ventral mesentery (vm), while the dorsal mesentery (dm) and the lateral mesentery (lm) are attached to muscle D. Trichrome, × 100.

Fig. 5   Cross-section at the point where all three mesenteries attach to the gut at the dorsal haemal vessel. Anterior to this point the gut is suspended in the dorsal mesentery (dm) as in figure 6. Posteriorly, the lateral mesentery (lm) is attached to the gut, while the ventral mesentery (vm) is attached to the dorsal mesentery as in figure 4. Trichrome, × 100.

of ciliated funnels, and the growing ventral mesentery is not attached to the gut (fig. 6). The participation of all three mesenteries is a transitory process, and by 72 hours the region of triple attachment has disappeared. The advancing ventral mesentery and the receding dorsal mesentery are attached to the gut at the mesenteric haemal vessel. The receding dorsal mesentery can be identified by the concentration of large nuclei along its edge (fig. 7). The advancing ventral mesentery is very thin and lacks the connective tissue layers of the dorsal mesentery. The lateral mesentery is no longer attached to the gut.

(2) *Posterior to the crossover point*

If the cut is made posterior to the point at which the digestive tract changes from stomach to intestine, all parts of the digestive tract are represented, to some extent, in the resulting anterior piece. The crossover point lies near the transection surface; the ventral mesentery is present and attached to the remaining short piece of intestine. To follow restoration of normal gut proportions in these regenerates, the body wall was marked at the stomach-intestine junction with vital stain. After 48 hours, the stomach-intestine junction, identified by the crossover, moves to a point between the stained mark on the body wall and the anterior end. This movement continues until the crossover point stabilizes near the midpoint of the animal, between 120 and 144 hours. The movement is accompanied by redifferentiation of the posterior part of the stomach to intestine.

(3) *The esophagus-stomach junction*

When transection is at the esophagus-stomach junction, both redifferentiation of the gut and axial elongation precede crossover. The time to crossover for half of a population of 20 animals was eight days. At this level, the actual time of

147

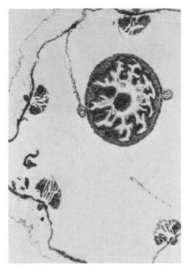

Fig. 6   Cross-section anterior to figures 4 and 5. Gut suspended in the dorsal mesentery (dm), no lateral mesentery (lm) present, and growing ventral mesentery (vm). Trichrome, × 100.

Fig. 7   The dorsal mesentery, showing the accumulation of cells present at the receding edge of the mesentery, 72 hours after section. No such accumulation of cells is seen in static mesenteries. Trichrome × 250.

TABLE 2

Time from transection to crossover at Level 3

Time	Experiments	Survivors	Crossovers
hours			
120	20	20	2
144	20	20	3
168	20	20	3
192	20	20	10
216	20	20	14
240	20	19	19

crossover is less predictable because the range within a population is quite large (table 2). By the time of crossover the gut has undergone extensive redifferentiation (fig. 8). Animals cut at this level showed a qualitatively obvious elongation. To test the hypothesis that a permanent growth region located in the anterior parts explains the regenerative ability of pieces containing the oral complex, a sliding scale was devised to measure elongation. The spicules of the body wall and the sensory papillae on the external surface provided natural markers. When used in conjunction with bands of vital stain, these markers showed clearly a region of rapid elongation just anterior to the amputation surface. Ten animals were marked by placing a strip of cellophane containing a vital stain on the body wall along the dorsal interradius. Then the animal was sectioned at the esophagus-stomach junction yielding an anterior piece marked with a stripe of stain, approximately 2 mm wide, and extending the length of the body. Between 120 and 192 hours, unstained, or lightly stained, tissues appeared in the stripe at the amputation surface, very close to, but not at, the posterior tip. The positions of the spicules and the sensory papillae in this light region were observed at 24 hour intervals. The positions of these markers with respect to each other changed daily in this region. By ten days, additional spicules and papillae were noted between the original, stained papillae. A few papillae at the posterior tip were still stained darkly, as were those anterior to the light region. Thin extensions of the longitudinal muscles could be seen through the

148

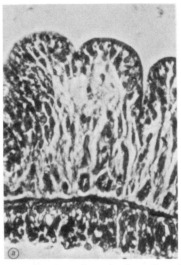 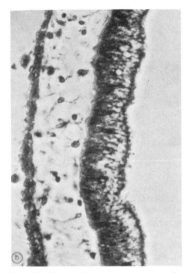

Fig. 8a Normal gut in the mid-esophagus region. The lumen is at the top. Note the radial connective tissue. The thick serosal epithelium is not a constant characteristic, but the lining epithelium is typical of the entire length of the esophagus. Trichrome, × 250.

Fig. 8b Gut of animal cut at the esophagus-stomach junction after four days of regeneration. This section is from the middle of the regenerate, well anterior to the region of rapid elongation. The lumen is to the right, the serosal side to the left. The lining epithelium has changed to resemble the normal epithelium of posterior gut regions. Note the regularity of the cell height, and the new organization of the connective tissue-fluid complex. Trichrome, × 250.

body wall in this region. No light tissues appeared in the stripe at the anterior end of the body during the ten days the experiment was continued. Clearly, the region of maximum elongation was just anterior to the posterior amputation surface. This observation does not imply the absence of growth throughout the animal, but it does show that an exclusively anterior growth region does not exist.

An alternate theory, that growth is accretionary with proliferative centers in the anterior end, was tested. Eight animals per day were studied histologically at two and four post-operative days. Six were sectioned at right angles, and two parallel, to the longitudinal axis. Every third section which passed through the nerve ring or the calcareous ring was checked for mitotic figures. These sections revealed a very small number of mitoses in some subdermal tissues. There were fewer than one mitotic figure for every five sections studied. No mitoses were found in the non-dermal components of the oral complex. Animals from a similar experiment fixed 12 days after operation showed a high incidence of mitosis in gut and body wall epithelia near the posterior end of the body where rapid elongation occurs.

DISCUSSION

Survival and regeneration of the anterior portion of specimens of *Leptosynapta crassipatina* divided into two parts across the body column is consistent and can be considered an established rule. These anterior pieces include the nerve ring, the calcareous ring, the mouth and the entire water vascular system as well as smaller structures such as the sense organs (Baur, 1864, Hamann, 1884) and the pulsatile rosettes (Becher, '07). Clark's suggestion (1899) that the reason

149

for the ability of anterior parts to survive and regenerate is the presence of the mouth and part of the digestive tract in these portions is untenable. First, regeneration proceeds at a reasonable rate in sterile, artificial sea water. In addition, the amount of digestive tract remaining in these smallest pieces of animal which can regenerate could not possibly be functional in assimilation of nutrients.

The process of regeneration which occurs after either autotomy or artificial section in *Leptosynapta* is the type which Morgan ('01) called morphallaxis; that is, the tissues which remain after injury are redistributed and remodeled to produce a functionally complete animal. This animal then grows to normal size.

One advantage of this type of regeneration is its speed. It allows restoration of function without a long and energetically expensive growing process. This is particularly important in *Leptosynapta* where the loss may include more than 95% of the length of the animal's digestive tract.

The histological observations reported here suggest that the initial event in crossover is the conjunction of the three mesenteries and the gut at the posterior amputation surface. Thus, as the ventral mesentery advances towards the anterior end of the animal, the dorsal and lateral mesenteries pull away from this attachment and attach to the surface of muscle D. It is the advancing ventral mesentery which pulls the gut across the coelom, and at the same time rotates the gut through 180°. The participation of the lateral mesentery if finished before 72 hours. This transitory participation is a recapitulation of the embryonic role of this mesentery (Runnstrom, '37). Evidently, the mesenteries play a significant role in these morphallactic movements. Previous workers have also found this to be the case. In most cases of regeneration of the viscera in echinoderms, the mesenteries are prominent. Torelle ('09), Kille ('36), Dawbin ('49a,b) and Reinschmidt ('69) have noted this fact previously in holothurian regeneration. Anderson ('65) dwells on the role of the mesenteries in visceral regeneration in asteroids. The processes of gut redifferentiation and crossover seem to be independent processes, since at some levels, redifferentiation precedes crossover.

The source of cells in the regeneration process has proven quite elusive, but several important points seem to have emerged from the work presented here. There is no proliferative center localized in the anterior region. Mitotic figures are not common at the anterior end. In addition, the region of maximum elongation is just anterior to the posterior amputation surface. Therefore, the viability of the anterior portion of *Leptosynapta* is not due to a permanent growth region at this level, nor is it due to a proliferative center which supplies cells to other parts. This growth pattern is analogous to the pattern in the asteroid starfish, with the main growth region lying just behind the advancing predifferentiated terminal tentacle and optic cushion.

Cell migration obviously plays a major role in regeneration in *Leptosynapta*. An increase in cell density at wound sites precedes the regenerative processes. The increase in cell density at the amputation surface is a very early indication of regeneration. Coelomocyte incorporation into regenerating structures has been suggested by several authors. Anderson ('62) proposed this cell source for the regenerating pyloric caeca of the asteroid *Henricia leviuscula*, but later retracted the suggestion (Anderson, '65). Cowden ('68) found no evidence for a role for coelomocytes in body wall regeneration in *Stichopus badionotes*.

There are three types of coelomocytes commonly found in *Leptosynapta* (Hetzel, '63, '65). These are the amoebocyte, the lymphocyte, and the morula cell. The morula cell has been implicated in production of connective tissue ground substance (Endean, '58; Doyle and McNiell, '64; Rollefsen, '65).

The clotting mechanism of the holothuroid *Stichopus californicus* is agglutination by amoebocytes in which the cells gradually lose their identity (Boolootian and Giese, '59). Clotting in *Leptosynapta* appears to be similar. Therefore, the accumulations of cells at the wound sites must contain large numbers of coelomocytes. The accumulation of cells in the edge of the receding dorsal mesentery probably consists of amoebocytes which are phagocytizing the connective tissues of the mesentery. However, there is no evidence of the participation of coelomocytes as components in the construction of new structures.

Even if coelomocytes are involved in regeneration, the stump tissues are also actively involved. There is significant degradation of muscle tissue at the amputation surface, and the degradation of other tissues may also occur. There is no clue as to whether the muscles contribute cells or just nutritive materials. The fragments found in the coelom are not nucleated.

In all cases of posterior regeneration, the tissues regenerated grow directly adjacent to stumps of similar tissues. No entirely different tissues are elaborated.

ACKNOWLEDGMENTS

This work is part of a thesis submitted to the Department of Biological Sciences, Florida State University, in partial fulfillment of the requirements for the Ph.D. degree. The work was directed by Dr. Michael J. Greenberg and completed in his laboratory. Initial support was provided by grant HE-09238 to Dr. Greenberg. The work was continued during the tenure of a NASA predoctoral fellowship and completed under U.S.P.H.S. predoctoral fellowship 1-F1-GM-38, 270-01.

LITERATURE CITED

Anderson, J. M. 1962 Studies on visceral regeneration in sea-stars. I. Regeneration of pyloric caeca in *Henricia leviuscula* (Stimpson). Biol. Bull., *122*: 321–342.
——— 1965 Studies on visceral regeneration in sea-stars. II. Regeneration of pyloric caeca in Asteriidae, with notes on the source of cells in regenerating organs. Biol. Bull., *128*: 1–23.
Baur, A. 1864 Beitrage zue Naturgeschichichte der *Synapta digitata* Nova Acta Acad. Leopoldina-Carolina., Bd. 31, 119 pp.
Becher, S. 1907 *Rhabdomolgus ruber* Keferstein und die Stammform der Holothurien. Z.f. wiss. Zool., *88*: 545–685.
Boolootian, R. A., and A. C. Giese 1959 Clotting of echinoderm coelomic fluid. J. Exp. Zool., *140*: 207–229.
Clark, H. L. 1899 The synaptas of the New England coast. Bull. of the U. S. Fish Comm., *19*: 21–31.

——— 1924 The Synaptinae. Bull. of the Museum of Comp. Zool., *65*: 459–501.
Cowden, R. R. 1968 Cytological and histochemical observations on connective tissue cells and cutaneous wound healing in the sea cucumber *Stichopus badionotus*. J. of Invert. Pathol., *10*: 151–159.
Dawbin, W. H. 1949a Auto-evisceration and regeneration of viscera in the holothuran, *Stichopus mollis*. Trans. Roy. New Zealand, *77*: 497–523.
——— 1949b Regeneration of the alimentary canal of *Stichopus mollis* across the mesenteric adhesion. Trans. Roy. Soc. New Zealand, *77*: 524–529.
Doyle, W. L., and G. F. McNiell 1964 The fine structure of the respiratory tree in Cucumaria. Quart. J. of Microscop. Sci., *105*: 7–11.
Endrea, R. 1958 The coelomocytes of *Holothura leucospilota*. Quart. J. Microscop. Sci., *99*: 47–60.
Fish, J. D. 1967 The digestive system of the holothurian, Cucumaria elongata. I. Structure of the gut and hemal system. Biol. Bull., *132*: 337–353.
Hale, L. J. 1964 Cell movements, cell divisions, and growth in the hydroid *Clytia johnstoni*. J. Embryol. and Exp. Morp., *12*: 517–538.
Hamann, O. 1884 *Beitrage zue Histologie der Echinodermen I. Die Holothurien*. Fischer Verlag. Jena, 529 pp.
Harvey, H. W. 1945 Recent Advances in the Chemistry and Biology of Sea Water. Cambridge University Press, London.
Heding, S. G. 1928 Papers from Dr. Th. Mortensen's Pacific Expedition 1914–16, XLVI, Synaptidae. Vidensk. Medd. Dank Naturhist. Foren., *85*: 105–323.
Hetzel, H. R. 1963 Studies on holothurian coelomocytes I. A survey of coelomocyte types. Biol. Bull., *125*: 289–301.
——— 1965 Studies on holothurian coelomocytes II. The origin of coelomocytes and the formation of brown bodies. Biol. Bull., *128*: 102–112.
Humason, G. L. 1967 Animal Tissue Techniques. second edition. W. H. Freeman and Company, San Francisco, 569 pp.
Kille, F. R. 1935 Regeneration in *Thyone Briareus*. Biol. Bull., *69*: 82–108.
——— 1936 Regeneration in holothurians. Yearbook. Carnegie Inst. Washington, *36*: 93–94.
Lillie, R. D. 1965 Histopathologic Technique and Practical Histochemistry. Third Edition. McGraw-Hill Book Co., Inc., New York, 715 pp.
Morgan, T. H. 1901 Regeneration. MacMillan, New York, 316 pp.
Pearse, A. S. 1909 Autotomy in holothurians. Biol. Bull., *18*: 42–49.
Reinschmidt, D. C. 1969 Regeneration in the sea cucumber, *Thyonella gemmata*. Masters Thesis, Florida State University.
Rollefsen, S. 1965 Studies on the mast cell-like morula cells of the holothurian *Stichopus tremulus*. Arbok for Universitetet I Bergen Mat. Naturv., Serie *8*: 1–12.

151

Runnstrom, S. 1937 Entwicklung von *Lepto-synapta inhaerens*. Bergens Museum Arbok 1.

Smith, G. I. 1964 Ed. Keys to Marine Invertebrate of the Woods Hole Region. Contribution No. 11, Systematic-Ecology Program, Marine Biological Laboratory, Woods Hole, Massachusetts.

Torelle, E. 1909 Regeneration in holuthuria. Zool. Anz., *35:* 15–22.

Vogt, W. 1925 Gestaltungsanalyse am Amphibienkeim mit ortlicher Vitalfarbung. Vorwarts uber Wege und Ziele. I. Methodik und Wirkungsweise der Vitalfarbung mit Agar als Farbtrager. Roux Arch. entwicks., *106:* 542–610.

Regeneration in the Sea Cucumber *Leptosynapta*
II. THE REGENERATIVE CAPACITY

GERALD N. SMITH, JR.

Members of the class Holothuroidea exhibit several patterns of autotomy and evisceration. The capacities of the separate parts to survive and regenerate are characteristic for each species, but a general pattern emerges. Species of the order Dendrochirota eviscerate through the anterior end and discard the calcareous ring, water ring, nerve ring, and pharynx, along with the other viscera (Pearse, '09; Scott, '14). The remaining parts—body wall, musculature, radial nerves and the cloaca—live and regenerate. In laboratory experiments, posterior thirds of *Thyone* (Torelle, '09; Kille, '35; '36) and *Thyonella* (Reinschmidt, '69) regenerate the entire animal; anterior pieces die. In the Aspidochirota, evisceration occurs through the posterior end (Domantay, '31; Bertolini, '30; Dawbin, '49a,b). The viscera are regenerated.

In addition, a cut within the body wall can be repaired, although any cut all the way through the body wall leads to total disintegration of the animal (Cowden, '68).

In those species of dendrochirotes which survive artificial section, the regenerative capacity is always associated with the posterior portion of the animal. In addition, the parts of the aspidochirotes capable of regenerating the entire animal are those retaining the cloaca. The one genus of the order Apoda which has been investigated presents a direct contrast to the Dendrochirota. In sea cucumbers of the genus *Leptosynapta*, survival capacity is restricted to the anterior end of the animal (Clark, 1899).

The surviving pieces regenerate quickly through a morphallactic process (Smith, '71). Any piece which includes the oral complex—mouth, pharynx, calcareous ring, nerve ring, water ring and water vascular system—lives and regenerates the rest of the animal. All other pieces die. These data suggest that some part of the oral complex is active in an essential physiological process; they do not prove any contribution of the oral complex to regeneration.

In this paper, the role of the oral complex in survival and regeneration is investigated through ablation experiments. Three primary questions are considered.

(1) What parts of the oral complex are essential for survival?

(2) What parts of the oral complex are essential for regeneration of a missing posterior half?

(3) Is there a part of the organism which can never be regenerated?

The procedure used was to excise a particular part and to observe the effect on survival. If the animal survived the excision of a part, the effect of this deletion on posterior regeneration was then tested. Finally, the ability of the animal to regenerate the excised part was checked histologically.

MATERIALS AND METHODS

Freshly collected specimens of *Leptosynapta crassipatina* from sand flats near the Florida State University Marine Laboratory at Turkey Point, Franklin County, Florida were used in all experiments. Survival experiments were carried out in artificial sea water (Harvey, '45) in glass aquaria or finger bowls at 21-23°C. The water was changed daily. Total inactivity and the beginning of autolysis were the criteria for death. Populations of experimental animals which survive for 12 postoperative days decrease, thereafter, at the same rate as controls. Therefore, for comparative purposes, survival capacities are reported here as the percentage of survivors remaining at day 12.

For long regeneration experiments the animals were kept on sea tables at the Florida State University marine laboratory at Alligator Harbor, Florida. They were supplied with running sea water, mud, and detritus. The progress of regeneration was monitored histologically. Animals were fixed in Bouin's, embedded in paraffin, and cut at 10 μ. The sections were stained with Masson's Trichrome connective tissue stain (Humason, '67).

In transverse section experiments, the animals were cut on paraffin backing with a razor blade. All other operations were performed in a dish of 0.54 M $MgSO_4$ on the transmitted light stage of a dissecting microscope. The animals were relaxed in $MgSO_4$ until the tentacles no longer contracted when touched. Tentacles were removed with micro-scissors. Pressure on the posterior half of the body column forces the tentacles straight out from the body, and keeps them extended during the operation.

Access to coelomic organs, and to the other parts of the oral complex, was possible through eversion of the animal. The eversion operation is performed in the following manner. The animal is relaxed and placed in the groove between the index and middle fingers of the left hand with the tentacles pointed towards the fingertips. The left thumb holds the animal in place. Utilizing a polished glass rod (2 mm by 75 mm), the tip of which is positioned on the buccal membrane, the tentacles and the entire oral complex are pushed back inside the coelom. The posterior end is cut from the body column, and the glass rod is pushed out through the resultant opening. This eversion process exposes the anterior gut, the water ring, and the posterior margin of the calcareous ring. Eversion can be maintained for up to an hour without ill effect.

Reversion is easily accomplished, but only if the animal has not been completely everted. While the coelomic lining is smooth and coated with a mucous slime, the spicules on the outside of the uneverted portion of the body wall stick tenaciously to the fingers and provide a firm grip. As a result, simply pulling the rod out is often sufficient to effect reversion. If necessary, however, the rod can be inserted through the posterior opening to accomplish reversion. Vigorous aeration speeds recovery and improves survival percentages. More than ninety per cent of the sham-operated animals in this study survived the entire eversion-reversion procedure.

Survival and regenerative capacity

Specimens of *Leptosynapta* were transected at various levels of the body column to test the ability of the resulting anterior and posterior parts to survive and regenerate. Five levels of section were tested (fig. 1). Two types of responses were observed. Animals cut at all levels anterior to the crossover point show similar survival curves. At Level I, more than 90% of the population lives for at least 20 days and begins regeneration. The posterior pieces decline rapidly, falling to 19% of the original population by the third day, and eventually leveling off near 6% by the tenth day (fig. 2). When section is posterior to the crossover point, a second type of curve is found (fig. 3). The initial mortality rate of posterior pieces is much lower, with 50% surviving for eight days. However, by day 12, only 15% still survive. At all levels the anterior and posterior curves are parallel after 12 days. The 12 day survival percentages for all five levels of section are given in table 1. By 12 days all anterior pieces show a positive regenerative response. No posterior piece shows any evidence of regeneration.

Levels of the oral complex

Anterior pieces of *Leptosynapta* include the oral complex—the mouth, pharynx, nerve ring, calcareous ring, and the water vascular system. While the oral complex is tightly organized, the region can be

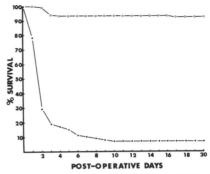

Fig. 2 Anterior vs. posterior survival for populations of animals cut at level I. Open circles, anteriors, dots, posteriors. Each population consisted of 120 pieces. Similar curves are obtained at levels II, III, and IV. Note that a small number of posteriors survived beyond 20 days. The final survivor from this population was fixed at 65 days.

subdivided by simple section. Three groups of 15 animals each were sectioned at different levels within the oral complex (fig. 4).

One group was sectioned at the posterior margin of the calcareous ring. The anterior-most pieces of the organism include the nerve ring, the calcareous ring and the tentacles. The water ring, stone canal, polian vesicle, and most of the tentacular ampullae were discarded along with the posterior parts.

A second group was prepared by removing the tentacles and sectioning posterior to the water ring. This leaves a short cylinder of tissue which includes the nerve ring, the calcareous ring, and most of the water vascular system. Only the tentacles are excluded.

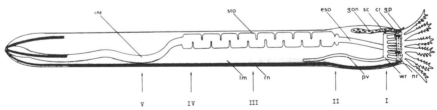

Fig. 1 Diagram of *Leptosynapta crassipatina* showing the five levels of section used in this study. The gonad and gonopodium are absent or much reduced in winter, but in summer the coelom is packed with branched gonad tubules. *Abbreviations:* int, intestine; sto, stomach; eso, esophagus; gon, gonad; sc, stone canal; cr, calcareous ring; gp, gonopodium; nerve ring; wr, water ring; pv, polian vesicle; rn, radial nerve; lm, longitudinal muscle.

155

TABLE 1

Survival of anterior and posterior halves of Leptosynapta sectioned at various levels

Level of section	Portion	Number of experiments	12 day survivors no.	%
Level I				
The water ring	Anterior	120	112	93.3
	Posterior	120	8	6.7
Level II				
The esophagus-stomach				
Junction	Anterior	25	25	100.0
	Posterior	25	2	8.0
Level III				
The middle of the stomach	Anterior	20	20	100.0
	Posterior	20	0	0.0
Level IV				
Anterior to the crossover				
Point	Anterior	20	20	100.0
	Posterior	20	1	5.0
Level V				
Posterior to the crossover				
Point	Anterior	40	38	95.0
	Posterior	40	6	15.0
Cumulative	Anterior	225	215	95.5
	Posterior	225	17	7.6

The third group was prepared by removing the tentacles and sectioning just posterior to the calcareous ring. The resultant oral disk contains the mouth, the calcareous ring and the nerve ring. These last named components of the oral complex are common to all three groups. The survival data are given in table 2. There is no significant difference in the survival capacities of the three groups; each showed good progress towards re-generation of the missing parts during the ten days of the experiment. Selected animals from group three were tested with a nutrient supply and showed complete regeneration and growth. Thus, the isolated oral disk including the nerve ring and the calcareous ring is sufficient for regeneration of the entire animal.

Lateral sections of the oral complex

Anterior pieces, produced by section at the posterior margin of the plates, were divided again, laterally, into various fractions of the oral complex. After 12 days, 17 of 18 halves were still living, and had

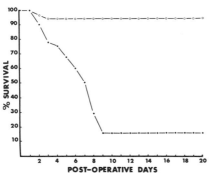

Fig. 3 Anterior vs. posterior survival for a population of animals cut at level V. Open circles, anteriors; dot, posteriors. Forty animals/group. Note high early survival of posteriors and stability of the population at ten days. The final survivor of the posterior population was fixed at 31 days.

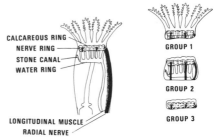

Fig. 4 Diagram showing pieces prepared by section within the oral complex. The nerve ring and calcareous ring are common to all three groups. All showed similar high survivals and all regenerated.

156

TABLE 2

Survival of pieces prepared by section within the oral complex

	Group 1	Group 2	Group 3
Experiments	15	15	15
12 day survivors	10	9	9
Regenerants	10	9	9

begun the deposition of calcium to form the missing half of the calcareous ring. These pieces included dorsal, ventral, and lateral halves; all failed at some point before regeneration of a complete animal. Histological sections showed that the nerve ring had regenerated prior to calcium deposition. Smaller fractions of the oral disk do not fare as well as halves, but all pieces begin an apparent regeneration process.

Half animals lacking selected organs

The ability of any surviving piece of the oral disk to "regenerate" prevents distinction between survival capacity and regenerative capacity in isolated pieces. However, a system to allow such a distinction can be devised in *Leptosynapta*. Posterior regeneration is a morphallatic process whereby the remaining tissues of the organism are redistributed and redifferentiated to produce a complete though smaller organism. This is best seen in animals cut at level IV. The posterior end of the stomach crosses from its dorsal position to the ventral side and redifferentiates to give intestine. The result is a functionally complete animal. This process is called crossover, and the stomach-intestine junction is called the crossover point (Smith, '71). Using the eversion technique and careful dissection,

it is possible to remove coelomic organs and parts of the oral complex from the animal. If, after reversion, the animal is cut just anterior to the crossover point, (Level IV), one can score survival, posterior regeneration and regeneration of the excised parts.

The following parts were removed from groups of 20 animals which were then cut at Level IV:

(1) the tentacles; (2) one half of the oral disk, including six tentacles and the associated nerve, water, and calcareous rings; (3) the esophagus; (4) the stone canal; (5) the stone canal and water ring, and (6) the entire oral complex.

A sham-operated control group was prepared, paired with group five, since these operations consume the most time. As each animal was everted to allow access to the stone canal and water ring, a second animal was everted and set aside in a dish of 0.54 mMgSO$_4$. On completion of the operation and reversion, the control animal was reverted and each pair of animals maintained in a divided culture dish.

The data (table 3) revealed a clear answer to the regeneration question. Crossover occurred in every case surviving four or more days, even those lacking the entire oral complex. Therefore, no component of the oral complex need be present for posterior regeneration to take place. The oral complex apparently cannot be regenerated by the rest of the organism, but it neither actively controls the process nor contributes to the process of regeneration.

The survival question is not answered so clearly. Half animals without the water ring seldom survive beyond 12 days

TABLE 3

Survival and regeneration of anterior halves lacking various parts

Parts removed from anterior halves	Number	12 day survivors		crossovers at four days
		no.	%	
Half rings	20	12	60	12 of 12
Tentacles	20	19	95	19 of 19
Esophagus	20	20	100	20 of 20
Stone canal	20	20	100	20 of 20
Stone canal and water ring	20	0	0	5 of 5
The oral complex	20	2	10	7 of 7
Controls	20	19	95	19 of 19

(table 2) and in no case regenerate a complete animal. But, an isolated oral disk without the water ring or stone canal lives and regenerates. To clarify the contribution of the water ring to survival in half animals, two additional experiments were performed. By sectioning the animal at the posterior margin of the calcareous ring it is possible to produce a posterior piece of *Leptosynapta* which retains the water ring. Twenty animals were transected at this level, and subsequently examined to verify the presence of the water ring and the stone canal in the posterior portion. All of the posterior portions were dead after 120 hours. Thus, the water ring and stone canal do not improve posterior survival when isolated from the remainder of the oral complex.

In the second experiment, 20 animals were everted and the connection between the water ring and the tentacles was cut without removing any tissue. After reversion, these animals also failed to survive beyond 120 hours. The entire control group of 20 animals survived. In this experiment, no tissue is removed from the organism. Only the connections between water ring and the tentacular ampullae are severed. This connection is thus essential to survival in organisms with a sizable coelom.

Regeneration of the excised parts

Surviving anterior halves of *Leptosynapta*, lacking selected portions of the oral complex, were able to regenerate posteriorly; they effected the crossover of the gut. The next question was whether these animals could regenerate the excised components of the oral complex. Accordingly, regeneration by the survivors of the operations listed in table 3 was observed. In addition, other experimental anterior halves were prepared. Selected animals were studied histologically, but no detailed histology of regeneration was attempted.

New tentacles regenerated from the stumps of the old. First, a conical extension grew outward from the stump, and then, after ten days, new digits began to appear in contralateral pairs. At 18 days, most tentacles were complete.

Regeneration of a lateral half of the nerve ring and calcareous ring, and of the six associated tentacles required special conditions. At 12 days, 12 such animals were still living (table 3). Histological study showed a regenerated nerve ring, but neither calcium deposition, nor tentacular outgrowth, had occurred. The experiment was repeated in running sea water with a supply of nutrient; complete regeneration occurred in 20 days. Presumptive nerve ring tissue, evident as a mass of histologically undifferentiated cells between the stumps of the old ring (fig. 5), appeared in animals fixed three and five days after ablation. At seven days, a complete nerve ring was present. After 16 days, tentacle stumps appeared and the calcareous ring had regenerated (fig. 6). The tentacles produced digits in 20 days. Otocysts, tentacle bases with ampullar valves, and pulsatile rosettes are present in the regenerated half of the oral complex.

The stone canal had regenerated by 18 days. A day 7 animal had an accumulation of histologically undifferentiated cells at the wound site. These cells presumably developed into the stone canal.

Parts of the gut regenerated readily after excision. The initial event was an accumulation of cells along the edge of the torn mesentery (fig. 7). This rod of histologically undifferentiated cells was then invaded by cells from the internal epithelial lining of the old gut (fig. 8). The invasion occurred from both anterior and posterior surfaces of the cut gut. The time course depended to some extent on the amount of gut ablated. A 2 cm piece excised from the anterior end of the stomach was regenerated, and had developed a patent lumen, in 96 hours.

Therefore, in every case in which deletion of a part was compatible with survival, the animal regenerated the excised part.

Posterior Regeneration

Posterior survival

Examination of survival curves of isolated posterior pieces (e.g., fig. 2) reveals that, even after 12 post-operative days,

158

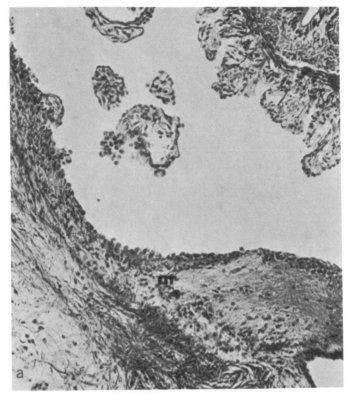

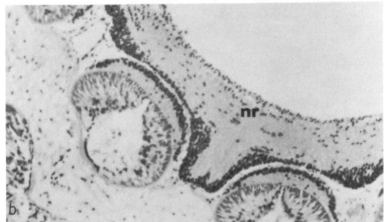

Fig. 5a   Regenerating nerve ring, (nr), five days after removal of half the oral complex. Trichrome × 250.

Fig. 5b   Normal nerve ring (nr) for comparison. Trichrome × 250.

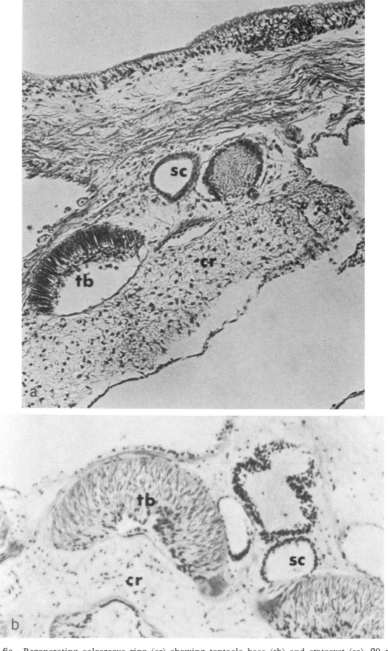

Fig. 6a   Regenerating calcareous ring (cr) showing tentacle base (tb) and statocyst (sc), 20 days after the operation. Trichrome × 250.
Fig. 6b   Control calcareous ring (cr), tentacle base (tb), and statocysts (sc). Trichrome × 250.

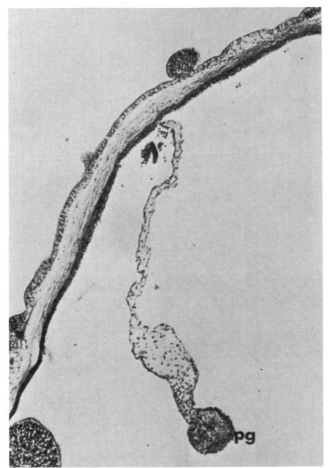

Fig. 7 Rod of presumptive gut (pg) cells along mesentery 48 hours after gut removal. Trichrome × 100.

a small percentage of the original population of pieces still survives. These survivors allow a test of regenerative ability of isolated posterior pieces. In the experiments which follow, the time periods used to test regenerative ability were limited by survival time periods. The procedure was to carry a population of posterior halves as long as possible; when only one or two individuals remained, the survivors were fixed and studied histologically.

Regeneration of posterior pieces in the anterior direction

Animals were bisected at one of three levels and the posteriors were observed for regeneration in the anterior direction. The levels of transection were Level I, just posterior to the water ring; Level IV, just anterior to the crossover point; and Level V, just posterior to the crossover point (fig. 2).

161

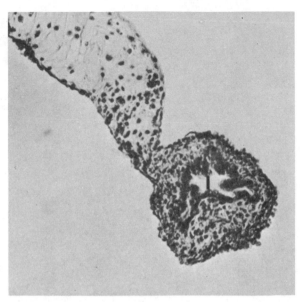

Fig. 8 Advancing lumen (1) of regenerating gut at more anterior levels of the animal shown in figure 7. Trichrome, × 250.

Animals sectioned at Level I were examined histologically for evidence of regeneration. More than 500 animals were cut at this level, but the total number of 12 day survivors was 21. Of these, two survived for 65 post-operative days. In no case did regeneration, beyond restoration of coelomic integrity, occur. Of nine cases in which survival exceeded 30 days, the ends of the radial nerves were partially reunited, but there was a gap across the dorsal interradius. No nervous tissue could be demonstrated crossing the dorsal mesentery. This mesentry was always present all the way to the new anterior tip. No degradation, or posteriad movement, of the mesentery was ever seen macroscopically or histologically.

When bisection was at the posterior end of the stomach, just anterior to the crossover point (Level IV), the result was similar. In one group of 60 experiments, there were seven posterior pieces which survived 12 days, and the final survivor was fixed at 28 days. The histological picture was similar to that at the water ring level. Also, the crossover point did not move toward the posterior. There was no evidence of gut redifferentiation.

The final level of bisection tested (Level V) was just posterior to the crossover point. Of 48 cases, six posteriors survived longer than 12 days; the final posterior was fixed at day 31. These posterior pieces had no dorsal mesentery, and no evidence of this tissue could be found, even after 31 days. The gut was typically intestine throughout its length; no morphallactic process was evident. That the radial nerves had failed to unite in two of these pieces was shown by histological examination.

To summarize, no evidence of regeneration was found at the anterior end of any posterior part of a transected animal.

DISCUSSION

The major result of this paper is the finding that the oral complex contributes nothing to posteriad regeneration. The regenerative capacity can now be stated in directional rules with the oral disk as the basic coordinate. Regeneration can occur in either direction away from the disk containing the nerve ring and the calcareous ring. It never occurs towards the oral disk.

No part of the oral complex imposes this unidirectional polarity on the system. In fact, the polarity holds even if the complex is absent; isolated posteriors undergo morphallaxis, and regeneration of missing, more posterior parts, quickly and without morphological error.

At any level, the tissues regenerate only those parts which are distal to the oral complex. Depending on the level of section, the amputation surface can lie anterior or posterior to a given portion of the digestive tract. If the amputation is posterior to a given portion of digestive tract, these tissues respond quickly to replace the missing parts of the gut. If the amputation occurs anterior to this portion of the digestive tract, the same tissues do not respond. None of the experiments discussed gives a clue to the nature of the unidirectional polarity of regeneration in *Leptosynapta*. All of the data support the notion of specific inhibition (Rose, '57) acting from posterior structures towards the anterior. Any part of the animal can be regenerated by remaining lateral parts of the organism, as shown for any half of the oral complex and for the gut.

The major role of the oral complex is its contribution to physiological maintenance. The continuity of the water vascular system is essential for survival of any part of the animal which includes a sizable coelom. Pieces of the oral complex lacking an extensive coelom are capable of regenerating the ampullae, the water ring, the polian vesicle, and the stone canal. But, when these parts are removed from any otherwise whole animal, the animal dies. These results suggest a role for the water vascular system in some essential physiological process. In a relatively constant environment, echinoderms face three basic problems: (1) assimilation of nutrients, (2) elimination of wastes, and (3) respiratory gas exchange.

The role of the mouth and pharynx in assimilation of nutrients is obvious. However, the following facts suggest that this function is not the critical role of the oral complex. First, regeneration takes place in sterile, artificial sea water (Smith, '71). Next, animals without digestive tracts survive and regenerate. Finally, *Leptosynapta* can absorb exogenous nutrient from sea water (Stephens and Schinske, '61). In our laboratory, even posterior pieces can incorporate radioactive amino acids and thymidine. Nevertheless, posterior pieces survive longer in artificial sea water than in running sea water with a high organic content. A medium containing one part glucose and 6.5 parts lactalbumin hydrolysate, or Eagle's medium, actually depresses survival. Survival improves as the concentration is lowered, but never exceeds control values (Smith, unpublished).

A role for the oral complex in waste elimination does not appear likely. The coelomocytes and ciliated funnels handle particulate wastes (Hetzel, '65). Regular changes of sea water probably insure removal of dissolved wastes.

The final possibility, that the oral complex is important in gas exchange, is promising. Ecologically, *Leptosynapta* occupies the interstitial niche. This area is characteristically low in oxygen. Runnstrom ('37) reported that *Leptosynapta inhaerens* lies in its burrow with tentacles exposed to the medium. The tentacles are thus in a favorable position for gas exchange. Tentacles are modified tube feet, and the tube feet of the sea urchin *Strongylocentrotus purpuratus* serve as the route of oxygenation (Farmanfarmaian, '59). Preliminary data indicate that raised oxygen tension can eliminate the initial high mortality of isolated posterior pieces between one and four days after sectioning (Smith, unpublished). If the pathway for gas exchange in *Leptosynapta* is through the tentacle surface to the fluid of the tentacular ampullae and the water ring, and thence to the coelomic fluid, the essential nature of these parts of the oral complex to survival, and thus to regeneration, becomes clear. An explanation for the shape of the survival curve for posteriors cut at the fifth level is possible on this basis. The body wall is very thin in these pieces, and oxygenation of the coelomic fluid may be effected by direct diffusion. Differing routes of oxygenation might lie behind the different survival capacities found among the holothurians. Those species in which the survival capacity is restricted to the posterior part of the animal utilize respiratory trees for respiration. These respiratory trees are outgrowths of the pumping cloaca, and the cloaca is always present in the parts of these animals which survive and regenerate.

ACKNOWLEDGMENT

This work was submitted to the Department of Biological Sciences, Florida State University, in partial fulfillment of the requirements for the Ph.D. degree. The work was directed by Dr. Michael J. Greenberg and completed in his laboratory. Support was provided by grant HE-09238 to Dr. Greenberg and by a NASA predoctoral traineeship and by U.S. P.H.S. predoctoral fellowship 1-F1-GM-38,270-01. Contribution 7 from the Tallahassee, Sopchoppy, and Gulf Coast Marine Biological Association.

LITERATURE CITED

Bertolini, F. 1930 Rigenerazione dell' apparato digerente nelle olothurie. Rend. Comp. Accad. Linei., 11: 600-601.

Clark, H. L. 1901 The synaptas of the New England coast. Bull. of the U. S. Fish. Comm., 19: 21-31.

Cowden, R. R. 1968 Cytological and histochemical observations on connective tissue cells and cutaneous wound healing in the sea cucumber Stichopus badionotus. J. of Invert. Pathol., 10: 151-159.

Dawbin, W. H. 1949a Auto-evisceration and regeneration of viscera in the holothuran, Stichopus mollis. Trans. Roy. Soc. New Zealand, 77: 497-523.

—— 1949b Regeneration of the alimentary canal of Stichopus mollis across a mesenteric adhesion. Trans. Roy. Soc. New Zealand, 77: 524-529.

Domantay, J. 1931 Autotomy in holothurians. Natural and Applied Sci. Bull. Univ. Philippines, 1: 389-404.

Farmanfarmaian, A. 1959 The respiratory surface of the purple sea urchin Strongylcentrotus purpuratus. Anat. Rec., 134: 561.

Harvey, H. W. 1945 Recent Advances in the Chemistry and Biology of Sea Water. Cambridge University Press, London.

Hetzel, H. R. 1965 Studies on holothurian coelomocytes II. The origin of coelomocytes and the formation of brown bodies. Biol. Bull., 128: 102-112.

Humason, G. L. 1967 Animal Tissue Techniques, 2nd Ed. W. H. Freeman and Company, San Francisco, 569 p.

Kille, F. R. 1935 Regeneration in Thyone briareus. Biol. Bull., 69: 82-108.

—— 1936 Regeneration in holothurians. Yearbook Carnegie Inst., Washington, 36: 93-94.

Pearse, A. S. 909 Autotomy in holothurians. Biol. Bull., 18: 42-49.

Reinschmidt, D. C. 1969 Regeneration in the sea cucumber, Thyonella gemmata. Masters Thesis, Florida State University.

Rose, S. M. 1963 Polarized control of regional structure in Tubularia. Develop. Biol., 7: 488-501.

Runnstrom, S. Entwicklung von Leptosynapta inhaerens. Bergens Museum Arbok 1.

Scott, J. W. 1914 Regeneration, variation, and correlation in Thyone. American Naturalist, 48: 280-307.

Smith, G. N., Jr. 1971 Regeneration in the Sea Cucumber Leptosynapta. I. The Process of Regeneration. J. Exp. Zool., 177: 319-330.

Stephens, G. C., and R. A. Schinske 1961 Uptake of amino acids by marine invertebrates. Limnol. Oceanogr., 6: 175-181.

Torelle, E. 1909 Regeneration in holothuria. Zool. Anz., 35: 15-22.

Origin of Calcified Tissue in Regenerating Spines of the Sea Urchin, *Strongylocentrotus purpuratus* (Stimpson): A Quantitative Radioautographic Study with Tritiated Thymidine [1]

The ability of sea urchins to regenerate damaged or lost portions of spines, and even entire spines, has been recognized for a long time and studies have been carried out in varying detail on this phenomenon by numerous investigators (see Heatfield, '71). Though the skeleton of regenerating sea urchin spines has received much attention, little work has been directed toward the soft tissues. During an investigation, by the author, of the histology and histochemistry of regenerating spines of the sea urchin, *Strongylocentrotus purpuratus*, questions arose regarding the origin of regenerated tissue. It was assumed that mitosis somewhere within the spine accounted for the appearance of new tissue, since it has been demonstrated that fractured spines also regenerate *in vitro* (Heatfield, '70, '71). In a radioautographic study, Holland ('65) described the pattern of incorporation of tritiated thymidine by whole spines of small specimens of *S. purpuratus* after incubation in the label for one hour *in vivo*. He concluded that mitosis occurs throughout the spine and is not restricted to certain zones such as the spine tip. However, Holland did not quantify his observations, and his conclusions may not be applicable to regenerating spines in which the pattern of cell proliferation might differ significantly from that in whole spines.

Reported here are the results of a quantitative radioautographic study undertaken with tritiated thymidine to establish the sites of cellular proliferation in regenerating spines of *S. purpura-*

[1] This work was supported in part by a National Aeronautics and Space Administration Predoctoral Traineeship and represents a portion of a doctoral dissertation submitted to the Department of Zoology, University of California, Los Angeles, California.

tus in order to elucidate the origin of regenerated tissue; specifically, the calcified dermis. The results of this study demonstrate the presence of a region of relatively high mitotic activity in the shaft of regenerating and whole spines at the level of the milled ring, and the absence of an apical growth zone. It is further shown that cells produced by mitosis in the shaft, migrate distally into the regenerate during growth of fractured spines and towards the tip of whole spines.

MATERIALS AND METHODS
Animals

Adult specimens of S. *purpuratus* ranging in wet weight from 40 to 60 gm were collected and maintained in the laboratory under constant conditions as described previously (Heatfield, '70).

Experimental procedures

Aboral, primary interambulacral spines of similar size were fractured with scissors two to three millimeters above the milled ring. The fractured stubs were then allowed to regenerate for various lengths of time *in vivo*. Regenerating stubs and non-fractured whole spines were then incubated under constant room lighting at 15°C in covered, plastic dishes with a capacity of 500 ml (Stoway utility dish, Southern California Plastic Co., Glendale) containing sea water and methyl-tritiated thymidine (^{3}HTdr) at a specific activity of 5000 mc/mM (Nuclear Chicago). The series of experiments employed in this study are summarized in table 1. Experiments I, II, and III were conducted *in vitro* with spines which had been removed (explanted) from each urchin by severing the tissue attaching the base to the underlying tubercle. Experiment IV was carried out entirely with spines incubated and maintained *in vivo*. Circulation of the incubation medium in experiments I, II, and III was achieved by directing a fine stream of air obliquely to the water surface. In IV, the medium was aerated via an air stone.

Histology

Labeled spines were fixed at room temperature for 24 hours or longer in neutral buffered, 10% formalin (Lillie, '65)

which was made approximately isosmotic with sea water (salinity of 34%) by the addition of sodium chloride. Fixed spines were decalcified for 72 hours in 10% ethylenediamine tetraacetate (EDTA), disodium salt, with a pH of 7.4 (Brain, '66). After washing in tap water for several hours, decalcified spines were dehydrated in methyl cellosolve (ethylene glycol mono-methyl ether, Union Carbide Corp.), three changes, 15 minutes each, followed by methyl cellosolve/methyl benzoate (1:1), 30 minutes. Tissues were then cleared in methyl benzoate, 24 hours, followed by fresh methyl benzoate, 24 to 72 hours, then three changes of benzene, ten minutes each. This was followed by benzene/paraffin (1:1), one hour. Tissues were then placed in three changes of paraffin, one hour each, and finally embedded in fresh paraffin (Tissuemat, mp of 56° to 58°C). Longitudinal serial sections of each spine were cut at 3 μ, and mounted on gelatinized, chromic acid-washed slides. Mounted sections were deparaffinized in toluene, three changes, two minutes each, transferred to absolute ethanol, two changes, two minutes each, and air dried prior to the application of liquid emulsion.

Histochemistry

Histochemical studies have demonstrated the presence of several types of migratory coelomocytes in spine tissues of S. *purpuratus* (Heatfield, unpublished). To histochemically differentiate these coelomocytes from connective tissue in the present work so that their contribution to the population of labeled cells could be assessed, a few slides were stained with the periodic acid-Schiff (PAS) technique of Lillie ('65) using the basic Fuchsin-Schiff reagent of Stowell ('45) prior to coating in liquid emulsion. After development, radioautograms which were prestained with PAS, were poststained with Naphthol yellow S according to the method of Deitch ('55).

To determine the extent to which tritiated thymidine is incorporated into DNA in these experiments, selected slides were deparaffinized, taken to water, and incubated at 37°C for periods up to 4.5 hours in 20 mM phosphate buffer (Lillie, '65), pH 7.0, containing 3 mM $MgSO_4$, and

TABLE 1

Summary of experiments with methyl-tritiated thymidine on cellular proliferation in regenerating and non-fractured, whole spines of S. purpuratus

Experiment	No. days of regeneration in vivo following fracture	No. hours of incubation in tritiated thymidine		No. days in plain sea water following incubation	Total no. days after fracture
		in vitro	in vivo		
I	12	1,2,5,11,26	—	—	13
II	11	24	—	—	12
	11 [1]	24	—	—	12
	(whole spines)	24	—	—	—
III	0	24	—	0,2,4	1,3,5
	3	24	—	0,2,4,6	4,6,8,10
IV	0	—	48	0,8,13,19	2,10,15,21
	(whole spines)	—	48	0,8,13,19	—

[1] Proximal one-half or more removed with scissors prior to incubation in tritiated thymidine.

0.2 mg/ml of electrophoretically purified, crystalline DNAase (Worthington Biochemicals), following the technique outlined by Barka and Anderson ('65). Control slides with tissue sections from the same spines, were similarly incubated in buffer, but without enzyme. All slides were then rinsed in 0.01 N HCl followed by distilled water. After drying, slides were coated with emulsion as described below. Developed radioautograms were left unstained, but mounted with a coverslip.

Radioautography

The procedures used are based largely on recommendations by Kopriwa and Leblond ('62). Darkroom operations were carried out under safelight illumination. Experimental and control slides of plain glass or with unlabeled tissue sections, were coated with the nuclear track emulsion, NTB-2 (Eastman Kodak), which was previously diluted 2:1 with distilled water. After drying in air, coated slides were stored at 4°C in light-proof boxes containing indicating Drierite. Exposure time varied from 3 to 17 days.

Exposed slides were developed for two minutes at 15°C in Kodak Dektol, diluted 1:1 with distilled water, then transferred to a stop-bath of distilled water for ten seconds, followed by Kodak acid fixer for three minutes, with a rinse and final wash in distilled water. After development, radioautograms were stained with Harris' hematoxylin and eosin (Humason, '67).

Counting of labeled nuclei

Nuclei were considered to be labeled in radioautograms if there were ten or more silver grains in the emulsion immediately above them on examination at × 970. Labeled nuclei in the dermis of calcified connective tissue were counted every 0.1 mm from the tip to the base of individual tissue sections in or near the median plane. In histological sections, the fracture plane can be easily distinguished by the abrupt change in tissue density along the spine axis. During sectioning and subsequent preparation of radioautograms, sections were frequently damaged. Thus, it was not always possible to count all of the labeled nuclei in several longitudinal sections in or near the median plane of each spine. Preliminary observations revealed that histograms of the axial distribution of labeled nuclei in five, undamaged serial sections from a single spine (example b of Day 4 in fig. 15), were virtually identical. As a result, it was decided to count labeled nuclei in a single, near-median longitudinal section selected from serial sections of each spine. The reliability of this method is reflected in the similarity of data obtained from duplicate spines assayed in most experiments. Occasionally, labeled nuclei were counted in portions of two non-adjacent sections of the same spine due to the difficulty of always obtaining a single, longitudinal section which passed through or close to the median plane from tip to base. Though labeling of nuclei occurred in the epidermis, these were not included in data presented below. The distribution of mitoses as established with ³HTdr in the present studies could not be correlated with the distribution of mitotic figures, since application of the Feulgen stain (Humason, '67) to selected tissue sections failed to demonstrate mitotic nuclei histochemically.

RESULTS

Spine anatomy and histology: general considerations

The spines of *S. purpuratus* are similar to those of other echinoids (see Hyman, '55) and consist of a fenestrated calcareous endoskeleton which is permeated throughout by a dermis of connective tissue containing migratory coelomocytes. The outer surface of the endoskeleton is covered by an epidermis which is continuous with that of the body wall. Longitudinally, the spine can be divided into a proximal base which is superficially separated from a distal tapering shaft by the milled ring (see Heatfield, '71). The concave base of the spine and rounded tubercle of the underlying body wall form a ball-and-socket joint. Muscle and connective tissue are attached to the body wall and along the base of the spine up to the milled ring (see figs. 2, 9). A detailed report of the histology, histochemistry and ultrastructure of spine tissues will be published elsewhere.

Incorporation of tritiated thymidine in vitro

Experiment I

To identify mitotic cells in spine tissues while regeneration is in progress, spines were explanted from an urchin 12 days after fracture, and incubated *in vitro* for periods up to 26 hours in 75 ml of sea water containing 1.0 μc/ml of ³HTdr.

Regenerating spines. Figure 1 shows the distribution of labeled nuclei along the axis of single sections from spines sampled after 1, 2, 5, 11, and 26 hours' incubation in ³HTdr. After one hour, labeled nuclei are distributed along the length of the regenerating spine, but are absent from the distal portion of the regenerate. It is apparent that labeled nu-

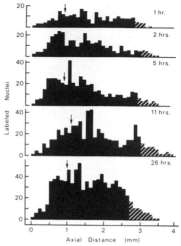

Fig. 1 Axial distribution of labeled nuclei in the calcified dermis in radioautograms of single, longitudinal histological sections of spines which were explanted 12 days after fracture and incubated up to 26 hours *in vitro* in tritiated thymidine. The horizontal base line in each example indicates the length of the calcified dermis in the section. Arrows indicate position of the milled ring. Labeled nuclei in the regenerate are represented by diagonal lines.

Fig. 2 Radioautogram of a longitudinal histological section of a spine which was explanted 12 days after fracture and incubated for 26 hours *in vitro* in tritiated thymidine (same section from which 26 hours' data in fig. 1 were obtained). Tissue-free spaces were originally occupied by the mineral component of the skeleton prior to decalcification in EDTA. Base, (b); epidermis, (ep); fracture plane, (fp); muscle, (m); milled ring, (mr); regenerate, (rg); shaft, (s). Scale line represents 0.5 mm. Formalin fixed; Hematoxylin-eosin stained; 3 μ.

Fig. 3 Increasing magnifications (3a–3c) of a portion of the regenerate indicated in figure 2. Only one nucleus is labeled (3c, arrow). Same magnification as figure 6.

Fig. 4 Increasing magnifications (4a–4c) of a portion of the regenerate indicated in figure 2. Arrows show some of the labeled nuclei. Same magnification as figure 6.

Fig. 5 Increasing magnifications (5a–5c) of a portion on both sides of the fracture plane indicated in figure 2. Arrows show some of the labeled nuclei. Same magnification as figure 6.

Fig. 6 Increasing magnifications (6a–6c) of a portion of the shaft of figure 2 near the milled ring. Arrows show some of the labeled nuclei. Scale line of 6a represents 100 μ; 6b, 20 μ.

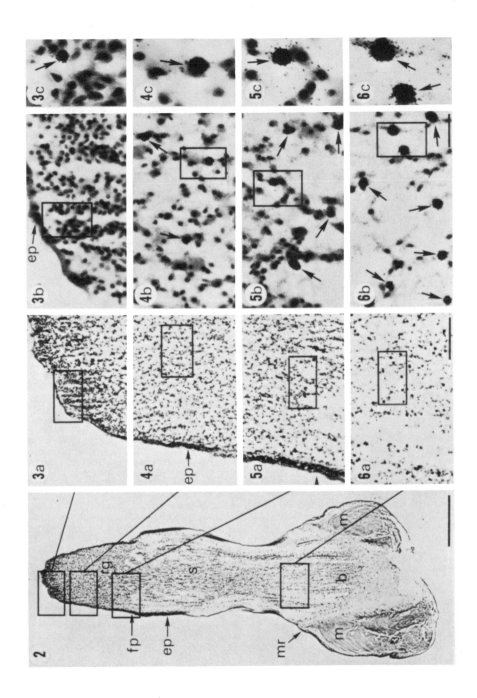

clei are distributed in a graded fashion with a maximum number occurring generally at the level of, or just distal to, the milled ring. With increasing periods of incubation, labeled nuclei appear more distally in the regenerate, with a few labeled nuclei present in the last 0.1 mm of the regenerate after 26 hours' incubation (see figs. 2–6). Similar results were obtained in a duplicate set of spines from the same urchin.

Figure 7 shows that the total number of nuclei in the calcified dermis (exclusive of coelomocytes) is highest in the regenerate, declining gradually to relatively low levels in the base. In comparison, the number of labeled nuclei reaches a maximum just distal to the milled ring after 26 hours' incubation in ³HTdr. There is a good correspondence between the distribution of labeled nuclei and the labeling index (percentage of cells incorporating the label), which reaches a maximum of 12.1% just distal to the milled ring. Significantly, the labeling index in the regenerate is relatively low, decreasing distally from 2.7% at the level of fracture to only 0.2% in the last 0.1 mm of the example shown. These data demonstrate that the relatively high mitotic activity in the vicinity of the milled ring after 26 hours' incubation in ³HTdr, is not due to a higher number of cells in this region, but reflects a non-uniform distribution in the proportion of nuclei incorporating the label.

The results of these experiments suggest that there is no localized, apical zone of proliferation at the growing tip of the regenerating spine of S. purpuratus. This leads to the hypothesis that cells, produced mitotically primarily in proximal regions of the shaft near the level of the milled ring, migrate distally to provide tissue for growth during regeneration. Consistent with this hypothesis are the following calculations. The total number of nuclei (exclusive of coelomocytes) counted in the calcified dermis of the regenerate in figure 7, 13 days after fracture was 6,336. The number of cells incorporating ³HTdr in the base and shaft (exclusive of the regenerate) during incubation for 26 hours was 845, or about 792 in 24 hours. Assuming that incorporation of the label represents cell

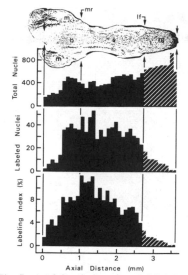

Fig. 7 Axial distribution of total (labeled plus unlabeled), labeled, and % labeled nuclei (labeling index) in the calcified dermis of the histological section (fig. 1, 26 hours, and fig. 2) shown in the photomicrograph. Labeled nuclei in the regenerate are represented by diagonal lines. Abbreviations as in figure 2; level of fracture, (lf).

division and that the rate of mitosis is relatively constant during regeneration in vivo, then a total of about 10,296 new cells would be produced during a postfracture period of 13 days. This value is in excess of the 6,336 nuclei counted in the regenerate of the tissue section, and suggests that cell division in the stump can adequately account for all of the cells required for growth of the calcified dermis of the regenerate, at least up to 13 days after fracture.

However, there is an alternative explanation for the absence of an apical growth zone in these experiments. If the epidermis is impermeable and the distribution of the label within the spine is limited by diffusion from the proximal end of explanted spines, the label might not have reached the tip of the regenerate in sufficient quantities even after 26 hours' incubation to be detectable by radioautography after incorporation into DNA. This is consistent with previous observations that labeled nuclei occur nearer the distal tip of the regenerate with increasing periods of incubation (see fig. 1). However, this latter observation might also be explained by migration of labeled cells, which would be consistent with the hypothesis proposed above to account for the existence of regenerated tissue.

Experiment II

To determine if the relative absence of labeled nuclei at the distal tip of explanted, regenerating spines is due to limited diffusion of the label from the proximal end, spines were explanted from a single urchin 11 days after fracture, and incubated *in vitro* for 24 hours in 200 ml of sea water containing 1.5 μc/ml of ^{3}HTdr. In some of the spines, the proximal one-half or more was removed by fracturing with scissors prior to incubation. This procedure removed that portion of the stub most active in incorporation and, hence, removal of the label, and reduced the distance the label would have to diffuse distally to reach the tip of the regenerate. In addition, whole spines were explanted from the same urchin and incubated with regenerating spines to determine the pattern of mitosis in nonfractured spines.

Effect of spine length. Figure 8 shows that the distribution of labeled nuclei in regenerating stubs from which the proximal portion had been removed is similar to that in comparable portions of intact regenerating stubs which served as controls. These results indicate that the relatively small number of labeled nuclei in the distal tip of the calcified dermis of the regenerate in previous experiments is not due to a diffusion barrier, but accurately reflects the relatively low level of mitosis in this region, and the absence of an apical growth zone. The distribution of labeled nuclei in the controls is similar to that observed in regenerating spines in experiment I (see fig. 1).

Non-fractured, whole spines. Figure 9 shows that labeled nuclei occur in the calcified dermis throughout the length of whole spines, with a maximum number near the level of the milled ring. In addition, there are relatively few labeled nuclei at the distal tip. This distribution of labeled nuclei is similar to that obtained for regenerating spines above, and

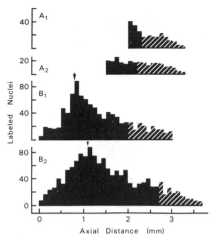

Fig. 8   Axial distribution of labeled nuclei in the calcified dermis in radioautograms of single, longitudinal histological sections of spines which were explanted 11 days after fracture and incubated for 24 hours *in vitro* in tritiated thymidine. In spines A_1 and A_2, the proximal one-half or more was removed prior to incubation in the label; B_1 and B_2 served as controls. Arrows indicate position of the milled ring. Labeled nuclei in the regenerate are represented by diagonal lines.

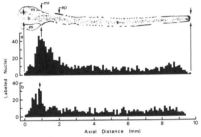

Fig. 9   Axial distribution of labeled nuclei in the calcified dermis in radioautograms of single, longitudinal histological sections of two non-fractured, whole spines which were explanted and incubated for 24 hours *in vitro* in tritiated thymidine. The photomicrograph shows the section from which the data in example (a) were obtained. Abbreviations as in figure 2. Formalin fixed; Hematoxylin-eosin stained; 3 μ.

suggests that a marked change in the distribution of mitoses does not occur in response to fracture and subsequent regeneration. However, the maximum number of labeled nuclei near the milled ring

171

in regenerating spines employed in this experiment is double that observed in whole spines (compare figs. 8, 9), suggesting that there may be an enhancement of proliferation rates during regeneration.

Pulse-labeling in vitro

Experiment III

To test the hypothesis that cells migrate distally during regeneration, spines were fractured on a single urchin. After regeneration for three days *in vivo* in plain sea water, additional spines were fractured on the same urchin. All stubs were then immediately explanted and incubated *in vitro* for 24 hours in 200 ml of sea water containing 1.5 μc/ml of ³HTdr. Thus, some stubs were incubated before, and some simultaneous with, the onset of regeneration of the calcified dermis (see Heatfield, '70). Some of the explants from each group were then fixed immediately, while the remainder were carefully rinsed, and placed in plain, aerated sea water for periods up to six days (see table 1, experiment III). All explants appeared healthy at the time of sampling.

Figure 10 shows the distribution of labeled nuclei in explants incubated *in vitro* in ³HTdr for 24 hours immediately after fracture, then maintained up to four days in plain sea water. The distribution of labeled nuclei in fractured spines fixed at the end of the incubation period (Day 1a, 1b of fig. 10) is similar to that found in comparable portions of regenerating and non-fractured spines in experiments I and II. However, by the fifth day after fracture (fourth day after incubation), there is a slight distal shift in the peak of the distribution of labeled cells. Significantly, a few labeled cells appear in the newly formed regenerate (see fig. 13) which was non-existent before and during incubation in the label (see fig. 11).

Figure 15 shows the distribution of labeled nuclei in explants incubated *in vitro* in ³HTdr for 24 hours after regeneration for three days *in vivo*, then maintained up to six days in plain sea water. The distribution of labeled nuclei in fractured spines fixed at the end of the incubation period (Day 4a, 4b of fig. 15) is again similar to that of comparable por-

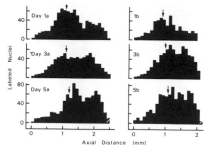

Fig. 10   Axial distribution of labeled nuclei in the calcified dermis in radioautograms of single, longitudinal histological sections of spines which were explanted immediately after fracture and incubated for 24 hours *in vitro* in tritiated thymidine, followed by plain sea water for periods up to four days. The number of days shown for duplicate spines indicates time after fracture. Arrows indicate position of the milled ring. Labeled nuclei in the regenerate are represented by diagonal lines.

tions of spines in experiments I and II. With increasing periods following exposure to the label, there is a conspicuous, distal shift in the peak of the distribution of labeled nuclei towards the fracture plane. This shift is interpreted as resulting from distal migration of labeled cells. It is noteworthy in this regard, that labeled nuclei occur throughout the calcified dermis of the regenerate (see fig. 14), which was in the early stages of growth during incubation in the label (see fig. 12). The total number of labeled nuclei in duplicate spines nearly doubled be-

Figs. 11–14   Portions of radioautograms of longitudinal histological sections of regenerating spines. The fracture plane is indicated by arrows. Scale lines represent 0.2 mm. Formalin fixed; Hematoxylin-eosin stained; 3 μ.

Fig. 11   Spine was explanted immediately after fracture and incubated for 24 hours *in vitro* in tritiated thymidine. Epidermis (ep) has nearly covered the fractured surface of the shaft.

Fig. 12   Spine was explanted three days after fracture and incubated for 24 hours *in vitro* in tritiated thymidine. Note that the regenerate is in the early stages of formation.

Fig. 13   Spine was explanted immediately after fracture and incubated for 24 hours *in vitro* in tritiated thymidine, followed by plain sea water for four days.

Fig. 14   Spine was explanted three days after fracture and incubated for 24 hours *in vitro* in tritiated thymidine, followed by plain sea water for six days.

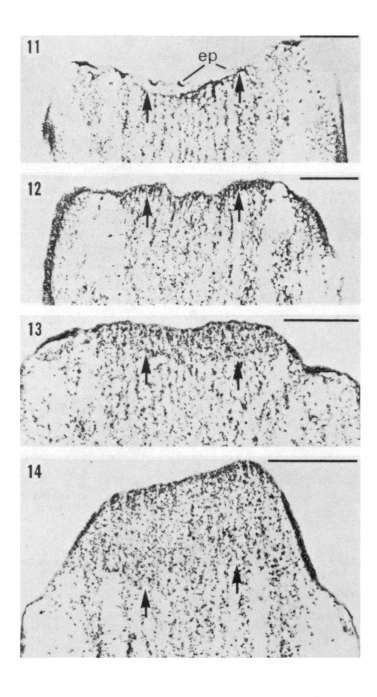

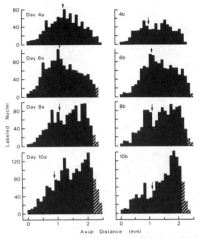

Fig. 15 Axial distribution of labeled nuclei in the calcified dermis in radioautograms of single, longitudinal sections of spines which were fractured and allowed to regenerate for three days *in vivo*, then explanted and incubated for 24 hours *in vitro* in tritiated thymidine, followed by plain sea water for periods up to six days. The number of days shown for duplicate spines indicates time after fracture. Arrows indicate position of the milled ring. Labeled nuclei in the regenerate are represented by diagonal lines.

tween day 4 and day 10, and there was a general reduction in labeling intensity with increasing time after incubation, suggesting that cells which had incorporated the label during incubation, subsequently divided.

The results of these *in vitro* pulse-labeling experiments are consistent with the hypothesis that cells produced by mitosis primarily in proximal regions of relatively high mitotic activity within the spine shaft near the level of the milled ring migrate distally during regeneration to provide tissue for growth.

Pulse-labeling *in vivo*

Experiment IV

To test the hypothesis of cell migration *in vivo*, spines were fractured and the urchin incubated for 48 hours in 300 ml of sea water containing 3.3 μc/ml of ³HTdr. After incubation, the urchin was rinsed and placed in plain sea water. Fractured and non-fractured spines were removed at intervals up to 19 days after incubation, and fixed (see table 1, experiment IV). Food material was provided during the post-incubation period, and the urchin appeared normal in all respects.

Regenerating spines. Figure 16 shows the distribution of labeled nuclei in spines incubated *in vivo* in ³HTdr for 48 hours immediately after fracture, then maintained *in vivo* up to 19 days in plain sea water. The distribution of labeled nuclei in fractured spines fixed at the end of the incubation period (Day 2a, 2b of fig. 16) is similar to that of comparable portions of explanted spines in previous experiments. However, with increasing periods following exposure to the label, there is a shift in the distribution of labeled nuclei toward the distal tip. After 19 days in plain sea water, a relatively large number of labeled nuclei occurs throughout the regenerate, which was non-existent before and during incubation in ³HTdr.

Figure 17 shows that during subsequent regeneration of fractured spines pulse labeled *in vivo*, there is a gradual increase in the proportion of labeled nuclei in the regenerate during growth, with a concomitant reduction in the shaft and base. In one spine (Day 21b of fig. 17), about 77% of the total population of labeled cells appears in the regenerate 19 days after incubation.

The results of these *in vivo* pulse-labeling experiments are consistent with the hypothesis that cells produced by mitosis primarily in proximal regions of relatively high mitotic activity within the spine shaft near the level of the milled ring migrate distally during regeneration to provide tissue for growth.

Non-fractured, whole spines. Figure 18 shows the distribution of labeled nuclei in non-fractured, whole spines incubated *in vivo* in ³HTdr for 48 hours, then maintained up to 19 days in plain sea water. The distribution of labeled nuclei in non-fractured spines fixed immediately after incubation in the label (Day 2) is similar to that of explanted, non-fractured spines in experiment II. With increasing periods following exposure to ³HTdr, however, there is a gradual distal shift in the distribution of labeled nuclei, with a reduction in labeled cells near

174

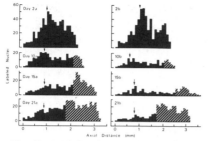

Fig. 16   Axial distribution of labeled nuclei in the calcified dermis in radioautograms of single, longitudinal histological sections of spines which were fractured and incubated for 48 hours *in vivo* in tritiated thymidine, followed by plain sea water for periods up to 19 days. The number of days shown for duplicate spines indicates time after fracture. Arrows indicate position of the milled ring. Labeled nuclei in the regenerate are represented by diagonal lines.

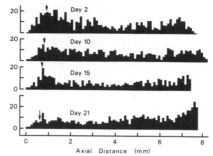

Fig. 18   Axial distribution of labeled nuclei in the calcified dermis in radioautograms of single, longitudinal histological sections of non-fractured, whole spines which were incubated for 48 hours *in vivo* in tritiated thymidine, followed by plain sea water for periods up to 19 days. The number of days shown for single spines indicates time after the beginning of incubation. Arrows indicate position of the milled ring.

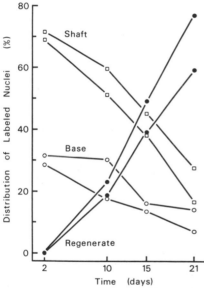

Fig. 17   The percentage of labeled nuclei in the base, shaft, and regenerate of duplicate spines with increasing time in days after fracture following exposure to tritiated thymidine. Data calculated from examples in figure 16.

the milled ring and a concomitant appearance of numerous labeled cells at the distal tip 19 days after incubation in the label. Similar results were obtained in a duplicate set of spines from the same urchin.

The maximum number of labeled nuclei near the milled ring in regenerating spines at the end of the incubation period in these *in vivo* experiments is double to triple the maximum number observed near the milled ring of whole spines. These observations are similar to those noted above for *in vitro* experiments, thus supporting the possibility that there may be an enhancement of the rate of cell division during regeneration, though more data are needed to conclusively establish this phenomenon.

Incorporation of tritiated thymidine by coelomocytes

Few instances of nuclear labeling of coelomocytes were observed in radioautograms prestained with PAS, and then poststained with Naphthol yellow S. It is concluded that the data presented in this paper largely reflects the incorporation of ³HTdr by cells of the calcified connective tissue of the dermis of the spine.

Localization of incorporated label

Labeled tissue sections digested with DNAase showed a marked reduction in labeling intensity compared to the controls. The reduction in labeling intensity is presumed due to removal of most, but not all of the labeled DNA by the action of DNAase. The relatively heavy labeling of nuclei did not permit a quantitative comparison between DNAase treated sections and controls. However, sufficient label was removed to conclude that most of the activity was incorporated into DNA. This conclusion is supported indirectly by the observation that silver grains were generally localized immediately over the nucleus of cells labeled with ^{3}HTdr (See Cleaver, '67, p. 37) in previous experiments.

DISCUSSION

Tritiated thymidine has been an extremely useful tool in the study of cell populations in animal tissues. It is incorporated specifically into DNA during replication prior to cell division, and allows subsequent study of growth, maturation, function, and death of the resultant labeled cells (see Cleaver, '67; Feinendegen, '67). In studies of regenerating tissues, the application of this technique can be particularly useful as a cell marker to elucidate the overall contribution of various cell types to the regeneration process, and can allow a determination of the origin of regenerated tissue. By labeling cells in this manner, studies on proliferation, migration, and changes in cytodifferentiation are made possible. In applying this approach quantitatively in the present investigation, considerable information has been obtained on the origin of regenerated tissue, and further, the results provide a dynamic picture of cell proliferation and migration during regeneration of the calcified dermis of experimentally fractured spines of S. purpuratus.

As elucidated in the present investigation and elsewhere (see below), the dynamics of growth during regeneration of endoskeletal tissues in fractured spines of S. purpuratus are in marked contrast to those found in similar studies on regeneration of endoskeletal tissues in other animals, namely the vertebrates, as exemplified by the limb of the amphibian, Triturus viridescens (see discussions by Hay, '66; Flickinger, '67; Schmidt,

'68; Goss, '69). Healing of the wounded surface of the spine shaft is completed by the second day after fracture, and consists of an inward migration of epidermis from the wound margin aided by division of some of the epidermal cells. By the third day after fracture, calcification is initiated in the regenerating dermis of the spine as evidenced by the simultaneous incorporation of calcium-45 into the skeleton (Heatfield, '70) and appearance of "microspines" on the fractured surface of the shaft (Heatfield, '71). In marked contrast to the regenerating limb of T. viridescens, there is as yet no evidence for dedifferentiation of existing tissues, formation and growth of a regeneration blastema, nor resorption of mineral preparatory to calcification in the regenerating spine. Regeneration of the lost portion of the spine shaft is usually complete about 60 days after fracture (Heatfield, '71). Regenerated tissue in the amputated limb of T. viridescens is ultimately derived from dedifferentiated, previously existing tissues of the old stump through proliferation and redifferentiation, while in spines of S. purpuratus, there is a continuous distal migration of new cells primarily from proximal regions of relatively high mitotic activity in the shaft quite distant from the actively growing tip of the regenerate. Thus, as new mineral is deposited apically and peripherally (Heatfield, '71), the spine grows and voids created in the regenerate proximal to the mineralizing front are filled by these migratory cells, and by migratory coelomocytes as well (Heatfield, unpublished). In this way the structural continuity of mineral and soft tissue of the spine shaft is maintained throughout the regenerative phase. Chalkley ('59) has shown a distal shift of the region of highest proliferation as regeneration of the limb of T. viridescens progresses. His observations were confirmed radioautographically by Hay and Fischman ('61). In the spine, a shift in centers of proliferation does not appear to take place during regeneration, though, as mentioned previously, the rate of mitosis may be elevated.

Based on the three major categories of cell populations that have been distinguished, i.e., stable, renewing, or expanding (see Leblond, Messier and Kopriwa, '59; Cleaver, '67), Holland ('65) concluded that S. *purpuratus* has strikingly few stable or renewing cell populations compared to those in the rat (Messier and Leblond, '60) and newt, *T. viridescens* (O'steen and Walker, '60). The only clearly renewing sea urchin cell populations that Holland observed were the tooth cells and spermatogenic cells of adult males during the ripe season, and the only static cell population in the adult urchin appeared to be the poison epithelium of the globiferous pedicellariae. He found that all other cell populations of S. *purpuratus* were expanding cell populations, though his experiments with whole spines of young urchins mentioned earlier were not designed to distinguish between renewing and expanding cell populations, nor did they shed any light on the dynamics of cell populations in regenerating spines or in mature spines which had ceased growth altogether. Therefore, one is tempted to state which of these categories best describes the dynamics of cell populations observed in spines of S. *purpuratus* studied here. If each of these categories is applied in a rigorous manner, then the results of the present investigation imply that during regeneration, and probably normal growth, the cell population is essentially an expanding one. However, the occurrence of a similar pattern of mitosis and cell migration in both non-fractured, whole spines and regenerating spines employed in the present work, raises the question of whether there is a continual proliferation and distal migration of cells throughout growth and development. Unfortunately, it is not known if non-fractured, whole spines studied here had ceased growth prior to incubation in ³HTdr. The possibility exists that they were in advanced stages of regeneration as a result of some previous damage or breakage. It is significant in this regard that spines of specimens of S. *purpuratus* observed in the field frequently show evidence of wear at the tips and may exhibit various stages of repair. Thus, it is not possible to draw conclusions about the long-term dynamics of the cell population of mature spines from the results obtained so far. If, as suggested by observations of animals in the field, the distal

tips of spines of S. *purpuratus* are continually being abraded so that the overall rate of growth (regeneration) is nearly balanced by the rate of wear, then the rigorous application of the concepts of a stable, renewing, or expanding cell population may not accurately describe the dynamics of the cell population, unless a comparison can be made with spines that have ceased growth. That spine growth is finite is suggested by recent observations (Heatfield, '71), and it may be possible to approach this problem experimentally. If proliferation and distal migration of cells continues once growth ceases, then they could be characterized as a renewing, rather than a stable population. Whatever the final outcome, it would appear that conclusions regarding cellular dynamics in spines of S. *purpuratus* should take into consideration the stage of development achieved at the time experiments are conducted and analyses are made.

ACKNOWLEDGMENTS

I would like to thank Dr. Dorothy F. Travis, Gerontology Research Center, Baltimore City Hospitals, Baltimore, Maryland, for kindly reading and commenting on the manuscript.

LITERATURE CITED

Barka, T., and P. J. Anderson 1965 Histochemistry. Harper and Row, New York, 660 pp.

Brain, E. B. 1966 The Preparation of Decalcified Sections. Thomas, Springfield, Ill., 266 pp.

Chalkley, D. T. 1959 The cellular basis of limb regeneration. In: Regeneration in Vertebrates, C. S. Thornton, ed. University of Chicago Press, Chicago, pp. 34–58.

Cleaver, J. E. 1967 Thymidine Metabolism and Cell Kinetics. John Wiley and Sons, New York, 259 pp.

Deitch, A. D. 1955 Microspectrophotometric study of the binding of the anionic dye, Naphthol Yellow S, by tissue sections and by purified proteins. Lab. Invest., 4: 324–351.

Feinendegen, L. E. 1967 Tritium-labeled Molecules in Biology and Medicine. Academic Press, New York, 430 pp.

Flickinger, R. A. 1967 Biochemical aspects of regeneration. In: The Biochemistry of Animal Development. Vol. II. Chapt. 6. R. Weber, ed. Academic Press, New York, pp. 303–337.

Goss, R. J. 1969 Principles of Regeneration. Academic Press, New York, 287 pp.

Hay, E. D., and D. A. Fischman 1961 Origin of the blastema in regenerating limbs of the newt *Triturus viridescens*. Develop. Biol., 3: 26–59.

———— 1966 Regeneration. Holt, Rinehart and Winston, New York, 148 pp.

Heatfield, B. M. 1970 Calcification in echinoderms: effects of temperature and Diamox on incorporation of calcium-45 *in vitro* by regenerating spines of *Strongylocentrotus purpuratus*. Biol. Bull., *139:* 151–163.

———— 1971 Growth of the calcareous skeleton during regeneration of spines of the sea urchin, *Strongylocentrotus purpuratus* (Stimpson): A light and scanning electron microscopic study. J. Morph., *134:* 57–90.

Holland, N. D. 1965 Cell proliferation in postembryonic specimens of the purple sea urchin (*Strongylocentrotus purpuratus*): an autoradiographic study employing tritiated thymidine. Ph.D. Dissertation, Stanford University, 224 pp.

Humason, G. L. 1967 Animal Tissue Techniques. W. H. Freeman, San Francisco, 569 pp.

Hyman, L. H. 1955 The Invertebrates. IV. Echinodermata. McGraw-Hill, New York, 763 pp.

Kopriwa, B. M., and C. P. Leblond 1962 Improvements in the coating technique of radioautography. J. Histochem. Cytochem., *10:* 269–284.

Leblond, C. P., B. Messier and B. Kopriwa 1959 Thymidine-H^3 as a tool for the investigation of the renewal of cell populations. Lab. Invest., *8:* 296–308.

Lillie, R. D. 1965 Histopathologic Technic and Practical Histochemistry. McGraw-Hill, New York, 715 pp.

Messier, B., and C. P. Leblond 1960 Cell proliferation and migration as revealed by radioautography after injection of thymidine-H^3 into male rats and mice. Am. J. Anat., *106:* 247–265.

O'steen, W. K., and B. E. Walker 1960 Radioautographic studies of regeneration in the common newt. I. Physiological regeneration. Anat. Rec., *137:* 501–509.

Schmidt, A. J. 1968 Cellular Biology of Vertebrate Regeneration and Repair. University of Chicago Press, Chicago, 420 pp.

Stowell, R. E. 1945 Feulgen reaction for thymonucleic acid. Stain Technol., *20:* 45–58.

178

THE EFFECT OF CESIUM-137 GAMMA RAYS ON REGENERATION IN *TUBULARIA*[1]

MARGARET SLAUGHTER, FLORENCE C. ROSE AND ANTHONY LIUZZI

MATERIALS AND METHODS

The forms of *Tubularia* used in these experiments grow in the Woods Hole, Massachusetts area throughout the summer. Only fresh material brought to the laboratory daily was used.

To induce the process of regeneration, stems 5 mm to 10 mm in length were cut from the colonies. A transverse cut was made below the hydranth and oblique cut 5 mm to 10 mm below the transverse cut. The transverse and oblique cuts were made to distinguish between distal and proximal ends of the cut stems. Cut stems were kept in glass finger bowls covered with cheese cloth and placed under running sand-filtered sea water.

Stems were irradiated using 667 Kev gamma rays from a 5000 Curie cesium-137 source at an exposure rate of 4000 R/min. All stems were irradiated in a Petri dish containing sea water and transferred to finger bowls postirradiation. Except for pilot experiments, all irradiations were completed within two hours after cutting.

[1] This work was supported in part by the U. S. Public Health Service Grant RH-70-01S469.

Exposures ranged from 10,000 R to 350,000 R requiring 2.5 to 87.5 minutes of irradiation.

The temperature during irradiation ranged from 10° C to 23° C. This was dependent upon the time required to accumulate a particular radiation exposure. For exposures greater than 75,000 R, iced sea water varying from 10° C to 18° C was used and changed during the irradiation procedure. Earlier observations showed that a temperature rise above 23° C during irradiation killed the organism.

Stems were observed under a dissecting microscope at different time periods post-irradiation. Seven stages of regeneration were defined. These were: pink, striated, double striated, pinched, bundle, emerging and emerged (Fig. 1). At

SEVEN REGENERATION STAGES

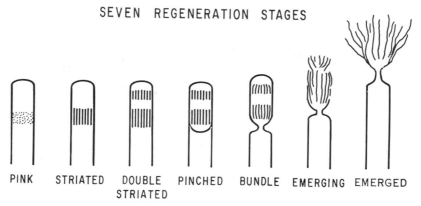

PINK   STRIATED   DOUBLE   PINCHED   BUNDLE   EMERGING   EMERGED
                    STRIATED

FIGURE 1. A diagrammatic representation of seven recognizable stages
in *Tubularia* regeneration.

each observation time a count was made of stems in each category. Stems undergoing no regeneration were also counted. Unirradiated stems fully regenerated complete hydranths within 36 to 48 hours after cutting. All stems died when kept more than five days in this particular laboratory situation.

RESULTS

Time of irradiation

Pilot experiments were performed to establish an optimum irradiation procedure. Preliminary experiments showed that stems of *Tubularia* will undergo complete regeneration at a radiation exposure of 10,000 R, but at a slower rate when compared with unirradiated stems. Experiments were performed to determine the most effective time to irradiate the organism. Groups of 10 stems were cut within the same time interval and exposed to 10,000 R at 2, 6, 12 and 18 hours after cutting. Comparisons were made of the number of stems from each group in the emerged stage of regeneration at these times. No stems, controls or irradiated, had reached the emerged stage 24 hours aftr cutting. Forty-two hours after cutting 0, 6, 3 and 9 stems irradiated, respectively, 2, 6, 12 and 18 hours

Table I

Accumulated delay in regeneration at various radiation exposure levels

# of stems in each group	Exposure	Time (hours)									
		30	36	42	48	54	60	66	72	78	90
		Per cent of stems pinched and beyond at various observation times after cutting									
20	Controls	25	55	85	80	95	100	100	100	100	100
20	10,000 R	10	40	60	80	85	90	90	85	80	80
20	20,000 R	5	30	40	50	65	80	90	85	85	85
20	50,000 R	0	5	5	30	30	40	65	65	70	70
20	75,000 R	0	0	0	0	5	10	20	45	45	65
20	100,000 R	0	5	5	5	15	20	25	25	40	40

after cutting had emerged. In addition, a group of 10 stems were irradiated at 10,000 R immediately after cutting; when observed 44 hours after cutting, only one stem in this group had emerged. Based on these data, it was concluded that maximum effectiveness is achieved if stems are irradiated within two hours after cutting. The general results presented below were based on subsequent experiments in which irradiation of stems was completed within two hours after cutting.

General radiation effects

Stems irradiated at 10,000 R and 20,000 R progressed through all the stages of regeneration, but the emerged hydranths had shorter tentacles than those hydranths regenerated from un-irradiated stems. This shortening of tentacle length became more pronounced as the radiation exposure increased.

At exposures of 75,000 R and 100,000 R no tentacles developed on those hydranths that formed, even though the hypostome emerged from the perisarc. Characteristic stages of *Tubularia* regeneration are identified by striations developing at the regenerating sites. The striations are developing distal and proximal

Table II

Delay in regeneration at various radiation exposure levels

# of stems in each group	Exposure	Time (hours)									
		20	36	42	48	54	60	66	72	78	90
		Per cent of stems emerged at various observation times after cutting									
20	Controls	0	1	35	50	80	90	100	100	100	100
20	10,000 R	0	0	5	35	55	90	90*	90	90	90
20	20,000 R	0	0	0	15	35	65	70	90*	90	90
20	50,000 R	0	0	0	0	20	25	50	40	65	70
20	75,000 R	0	0	0	0	0	5	15	25	40	65
20	100,000 R	0	0	0	0	5	10	20	25	35	40

* Indicates that at later observations emerged hydranths had begun to drop off.

tentacles. The lack of tentacle growth at high exposures made the various stages of regeneration less recognizable.

Observations and comparisons of regenerated hydranths showed that hydranths regenerated from irradiated stems tended to drop off sooner than those hydranths regenerated from un-irradiated stems.

Stems were exposed to 50,000; 75,000; 100,000; 150,000 and 200,000 R to determine an exposure that completely inhibits regeneration. Observations of the stems after irradiation indicated that 150,000 R and 200,000 R inhibit regeneration, but were not lethal to the organism 90 hours after cutting.

The inhibitory effect of irradiation appeared to be dose dependent. As the exposure increased, the delay in the time of regeneration increased. From recordings of the number of stems in each stage of regeneration for each exposure, all stages appeared to be uniformly affected by irradiation. Table I is a summary of data on delay in regeneration at various radiation exposure levels based on the percentage of stems completing the pinched and all subsequent regeneration stages within the given observation intervals. Table II summarizes data on the observed delay in regeneration time of emerged stems at various exposure levels. In this experiment those irradiated stems that did emerge were, respectively, 6, 12, 18 and 30 hours behind the un-irradiated stems. Figure 2 is a graphic representation of the data in Table II. The percentage of stems completing the emerged stage of regeneration decreases as the exposure levels increase.

Stems were exposed to radiation levels ranging from 225,000 R to 350,000 R in steps of 25,000 R. At an exposure of 300,000 R, 10 out of 20 stems showed

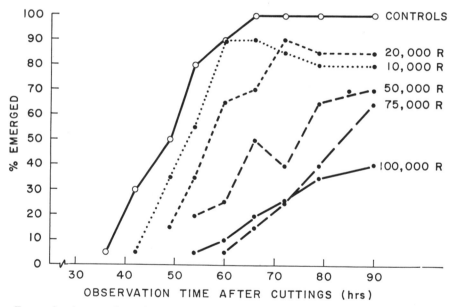

FIGURE 2. A comparison of the percentage of irradiated and control stems in the emerged stage of regeneration at different observation times after cutting.

signs of tissue disintegration when observed six hours after cutting. At eighteen hours after cutting and irradiating all 20 stems were dead. Exposures of 300,000 R and above are fatal to the organism under the given experimental conditions. For the various experiments described a total of 900 organisms was used in ten distinct experiments.

DISCUSSION

It is well-known that ionizing radiation affects regeneration in various invertebrates and vertebrates. Irradiation has frequently been used as a means of studying biological mechanisms involved in the regeneration process. The quantity of radiation necessary to inhibit regeneration is extremely variable. In the coelenterates, various genera react differently; some coelenterates are more radiation resistant that others. X-irradiated hydra whose interstitial cells have been destroyed are still capable of limited regeneration. (Brien and Reniers-Docoen, 1955). Also for the hydra, an exposure of 4500 R inhibits bud formation (Park, 1958). Colonies of *Pennaria tiarella* require 10,500 R to inhibit regeneration. For the colonial hydroid, *Tubularia,* an exposure of 150,000 R is required to inhibit regeneration.

Inhibitory exposures have also been determined in various other organisms. Regeneration in the polychaete, *Clymenella torquata* is inhibited at an exposure of 50,000 R (Rose, F., unpublished observations). X-irradiation ranging from 1000 R to 10,000 R when applied locally will prevent limb and tail regeneration in the adult urodele, *Triturus* (reviewed by Rose, 1964).

The lethal exposure of 300,000 R observed in the present study for *Tubularia* is in the range of exposures required to deactivate many viruses. Viruses lose their infectivity when exposed to radiations ranging from 430,000 R to 5,150,000 R (Luria, 1951).

The authors would like to express their sincere appreciation to Dr. S. Meryl Rose for his advice and interest in this investigation.

LITERATURE CITED

BARDEEN, C. R., AND F. H. BAETJER, 1904. The inhibitive action of the Roentgen rays on regeneration in planarians. *J. Exp. Zool.,* 1 : 191–195.

BRIEN, P., AND M. RENIERS-DECOEN, 1955. Cellules interstitielles des hydres d'eau douce. *Bull. Biol. France Belq.,* **89** : 259–325.

BUTLER, E. G., 1931. X-radiation and regeneration in *Amblystoma. Science,* **74** : 100.

CURTIS, W. C., AND R. RITTER, 1927. Further studies on the effects of X-radiation on regeneration (abstract). *Anat. Rec.,* **37** : 128.

LURIA, S. E., 1951. Radiation and viruses, pp. 333–364. *In:* Alexander Hollaender, Ed., *Radiation Biology, Volume II.* McGraw-Hill- New York.

PARK, H. D., 1958. Sensitivity of hydra tissues to X-rays. *Physiol. Zool.,* **31** : 188.

PUCKETT, W. O., 1936. The effects of X-irradiation on the regeneration of the hydroid, *Pennaria tiarella. Biol. Bull.,* **70** : 392–399.

ROSE, S. M., 1964. Regeneration, pp. 545–622. *In:* J. A. Moore, Ed., *Physiology of the Amphibia.* Academic Press, New York.

KEY-WORD TITLE INDEX

AUTHOR INDEX